MEDICINE, SCIENCE, AND RELIGION
IN HISTORICAL CONTEXT

Ronald L. Numbers, *Consulting Editor*

Christian Science on Trial

Christian Science
on Trial

Religious Healing in America

Rennie B. Schoepflin

THE JOHNS HOPKINS UNIVERSITY PRESS
Baltimore & London

© 2003 The Johns Hopkins University Press
All rights reserved. Published 2003
Printed in the United States of America on acid-free paper
2 4 6 8 9 7 5 3 1

The Johns Hopkins University Press
2715 North Charles Street
Baltimore, Maryland 21218-4363
www.press.jhu.edu

Library of Congress Cataloging-in-Publication Data
Schoepflin, Rennie B.
Christian Science on trial : religious healing in America / Rennie B. Schoepflin.
p. cm. — (Medicine, science, and religion in historical context)
Includes bibliographical references and index.
ISBN 0-8018-7057-7 (alk. paper)
1. Christian Science—History. 2. Medicine—Religious
aspects—Christian Science—History. 3. Medical care—Law and
legislation—United States. I. Title. II. Series.
BX6950 .S34 2002
289.5′73—dc21
2001008512

A catalog record for this book is available from the British Library.

For Charles

Contents

Acknowledgments

Umberto Eco tells us that when Marco Polo traveled to China he looked for unicorns. He found them when he saw rhinoceroses. So much of what we see is determined by our expectations; so much of what we know is a reflection of our background knowledge. Trying to see Christian Science healers, so familiar yet so foreign, often made me feel like Marco Polo. I owe a debt to those who encouraged me to look beyond my expectations and yet to translate what I saw into the familiar.

As this project evolved over the years, several published papers, numerous public presentations, and wide-ranging conversations provoked my thinking and ripened my understanding. Among the many, I owe much to Judith Walzer Leavitt, Norman Gevitz, Rima D. Apple, Robert David Thomas, Jonathan M. Butler, Robert Peel, Stephen R. Howard, and Darrel W. Amundsen. And I am especially indebted to the advice, support, and friendship of Ronald L. Numbers.

My thanks to the staff in numerous municipal, county, and state courthouses across the country who helpfully pointed me toward the dusty trial records that were essential to my study. Librarians and archivists at many institutions helped me with my research, but in particular I acknowledge the La Sierra University librarians and the archivists at Boston's Christian Science Center, Duke University Library, and the Phillips Memorial Library at Providence College.

Colleagues, former student assistants, and my university have shown sustained interest in and support of the project. In particular I thank Sally Andriamiarisoa, James Beach, Greg Cushman, Clark Davis, and David Pendleton.

Editor Jacqueline Wehmueller, anonymous reviewers, and the staff at the Johns Hopkins University Press offered advice at crucial times and never lost faith.

Finally, with gratitude I acknowledge the support and encouragement of family and friends—Le Ann, Charles, Sally, and Jim.

Christian Science on Trial

Introduction

Recently the United States has experienced a resurgence of interest in spirituality and health, from crystal therapy and shamanic healing to therapeutic touch, spiritual massage, and charismatic healing. Deepak Chopra's Center for Well Being thrives in La Jolla and in cyberspace, Andrew Weil's program in integrative medicine prospers in Tucson and on *Larry King Live*, and Herbert Benson's mind-body medicine flourishes in supposedly buttoned-down Boston. Many contemporary Americans seek to restore and sustain physical well-being through ancient or New Age healing practices based on a spiritual understanding of the universe. Even old-fashioned prayer to heal the sick seems to be alive and well in this postmodern age. One recent poll found that 82 percent of Americans believed in the healing power of prayer and 64 percent thought doctors should pray with patients who request it.[1]

This interest in spiritual healing appears to be more than a fad; research institutes affiliated with big-name universities have sprung up across the country to study or exploit it, and the National Institutes of Health (NIH) spend tax dollars to discover if spiritual healing works. The Center for the Study of Religion / Spirituality and Health at Duke University explores how religious and spiritual beliefs and practices affect physical and mental health; the Mind / Body Medical Institute, affiliated with Harvard Medical School and Beth Israel–Deaconess Medical Center, regularly conducts conferences on spirituality and healing; and the program in integrative medicine at the University of Arizona embraces a "comprehensive approach to the art and science of medicine" that includes the spiritual and the transcendent.[2] The International Center for the Integration of Health and Spirituality (formerly the National Institute for Healthcare Research), which is "dedicated to the integration of health and spirituality" and to the advance of "this emerging field," runs a thriving operation.[3]

The NIH's National Center for Complementary and Alternative Medicine has awarded grants to investigators seeking to document the medical efficacy of a wide range of spiritual practices, including so-called bioenergy theory, spiritual and energy healing, and prayer at a distance. Whether these investigations will corroborate the studies executed at McGill University and Duke University Medical Center that suggested the efficacy of prayer remains to be seen. At least they will add to the nearly fifteen hundred research studies, research reviews, articles, and clinical trials that have appeared on the subjects of spirituality/religion and medicine/health since 1990.[4]

This American revival of traditional and New Age religious healing in the late twentieth century bears an uncanny resemblance to the emergence at the turn of the nineteenth century of mind healing, religious healing, New Thought healing, and most notably, Christian Science healing. Just at a time when American physicians increasingly advocated a model of health that segregated the mind and body, relegating the former to the church or psychotherapeutics while reserving the latter for medical science, Christian Scientists promoted a radical spiritualization of health that merged body into mind, rendering religion and science one. When we examine the American medical community's reactions to spiritual healing in each century, however, the parallels break down. While many of today's physicians sneer at the contemporary rise of spiritual healing and decry its supposed threat to the health of children, fewer respond with the animosity apparent at the turn of the century. This different reaction reflects both the enhanced status of scientific medicine and the marginal contribution of Christian Science to this late twentieth-century revival of spiritual healing.[5]

This book presents for the first time a detailed analysis of the structure and content of the turn of the nineteenth century legislative and legal confrontations between organized medicine and Christian Scientists. That analysis in turn uncovers a new understanding of both Christian Scientists and the medical community. The anxiety and tenacity physicians exhibited as they fought to trivialize and curtail Christian Scientists challenge the sunny assumptions made by many historians regarding the confidence of the medical profession about who would control the future of American health care. At the time, scientific medicine and religious healing were locked in what many considered a life or death struggle for ascendancy, whose outcome seemed uncertain. Early victories in attempts to curtail the practice of religious healers made it appear that scientific medicine had won the conflict. A closer reading of the events and the advantages of hindsight, however, make it seem more likely that a relationship of complexity or convergence had evolved between spiritual healing and medical practice.[6]

The limited authority of the medical community became notably transparent in the pitched battles fought by physicians, Scientists, and other interested parties in America's courtrooms and legislative halls over the legality of Christian Science healing. Initially those complex conflicts wove together the issues of medical licensing, the meaning of medical practice, and the supposed right of Americans to therapeutic choice. Then the legal issues shifted to contagious disease, public safety, and children's rights. But the disputes also accentuated the gender, attitudes, and activities of Christian Scientists. Why were there so many women in the movement? Did they claim to practice medicine? Did they diagnose contagious diseases? Would they submit their children to the care of physicians? In each case the answers surprised Americans.[7]

Key members of the medical community moved to exclude Christian Science practitioners from the medical landscape by promoting a state by state effort to restrict their practice through legislation. As doctors considered what the apparent successes of Christian Science healing meant for their own practice, they adopted or subsumed what they thought worked into psychotherapeutics. When Christian Scientists effectively fended off the medical onslaught, it spoke well of their political lobbying efforts and their willingness to make timely accommodation, but their legislative successes also bespoke a public support of religious healing and underscored that American physicians lacked a strong position of cultural authority over health care until after the 1920s.

For a time Scientists, like chiropractors and osteopaths, attempted to have legislatures and courts recognize their method of healing as just another medical sect that American patients should be free to choose. But Scientists came to doubt such a strategy, acknowledging that the licensing procedures such an approach required would corrode the spiritual essence of their healing, and they relinquished their claim to practice medicine in the open market. Instead they lobbied for legislation that would carve out an unmolested niche for believers to practice religious healing exclusively on themselves and their children. Again, however, they ran head-on into a medical community outraged by the public health dangers of "untreated" contagious diseases and convinced that medicine could save sick children while Christian Science could not. In a series of accommodations to quarantine and vaccination laws and to laws requiring the reporting of contagious disease, Christian Scientists continued publicly to adjust their practices to a hostile world while working quietly behind the scenes for legislative relief. The accommodations to a modern world forced upon them in the early twentieth century and the untiring efforts of the church's Committees on Publication to fashion legislation that suited their members' needs worked well for much of the century to keep the peculiar beliefs and practices of Christian Scientists on the periphery of public consciousness and safe from legislative intrusions. But as the twentieth century drew to a close, old conflicts with physicians and prosecutors arose to bedevil the movement anew. A national campaign against child abuse and the wrenching deaths of Christian Scientist children spawned a move to prosecute parents for child endangerment when they chose spiritual healing for their children.

Here I argue that the court trials of Christian Science practitioners and parents between the mid-1880s and the 1920s and again during the 1980s were only in part about popularly understood notions of law, guilt, and innocence. The trials also exposed the competing ways the rhetorical forms that prevailed inside a courtroom could be used to reconstruct the story of American health care and thereby influence its future on the outside.[8] Recent scholars have clarified the central role that rhetoric plays in the courtroom; despite all the rules that circumscribe it, a trial is not a mechanical process but "a set of rules for reconstructing a disputed incident in a symbolic form that allows the actions of the participants to be judged in fairly uniform fashion by [someone] who was not witness to the incident." A symbolic form, the story, integrates the whole; but the power of storytelling depends on cognitive and communication skills. Individuals or groups may suffer injustice "simply because they fail to share the communication and thought styles used by dominant segments of the population."[9]

Prosecutors repeatedly questioned Scientists to elicit responses that highlighted their idiosyncratic views of the body and of symptoms, sickness, death, and prayer. A judge or jury who did not share, let alone understand, the Scientists' worldview often found it difficult if not impossible to understand their story, and judicial injustice was a distinct possibility. Viewed in this way, the marginal worldview of Christian Scientists may explain why some believed the courts unjustly judged or even persecuted them. But physicians often found themselves mistrusted as well. Defense attorneys stressed the threat of medical monopoly and derided the "imaginary" germs, inflated cure rates, and underreported death rates of scientific medicine. The resultant ambiguity of these court decisions before World War I reflected in part the absence of a coherent and commonly held American view of medicine and health care.[10]

Although Christian Scientists and physicians took center stage in these dramas, attorneys, clergy, legislators, and the press played significant minor roles, and the American public stood like a Greek chorus, bringing the sovereign judgment of the republic to the debates and their issues. As popular, political trials, they allowed competing groups within society to use the courts to disadvantage their rivals in a power struggle.[11] But more important, they attracted the attention of a wide audience, often through extensive press coverage, and played a distinctive role in the formation of an American consensus about scientific medicine that in turn defined and transformed Christian Science. The popular nature of the trials allowed them to function as forums for public debate and provided the public with explicitly public and authoritative norms for resolving their disputes about medical care.[12]

The decades of sustained engagement fundamentally altered the beliefs and practices of both Scientists and physicians, but physicians' public victories became emblematic of the way scientific medical practice came to publicly dominate spiritual healing in America for much of the twentieth century. Only in the last twenty-five years of the century did a resurgence of interest in the spiritual dimensions of human health give reason to believe that the apparent public hegemony of medical science over religion might be weakened, if not broken.[13] And Christian Scientists, with their clarion call for spiritual healing in a materialist age, found themselves once again at center stage.

To better ground our perspective on these turn-of-the-century clashes over Christian Science healing, this book develops an understanding of Christian Science as a medicoreligious movement born out of the personal experiences of its founder, Mary Baker Eddy. The recurring illnesses, unfulfilled expectations, marital and financial problems, and conflicts with students and former healing partners that filled her early years exposed a motif of trial and tribulation that foreshadowed the history of Christian Scientists. Throughout her evolution as a healer and teacher, Eddy exhibited the classic profile of a spiritual seeker, searching for physical and mental health and personal autonomy and "discovering" them through spiritual enlightenment.

Christian Scientists traced their religious roots to 1866, when Mary Baker Eddy spontaneously recovered from a severe injury that authenticated her "discovery" that reality is completely spiritual and evil is only an illusion. Scientists believed that Eddy had recovered the cardinal teaching of Jesus Christ that all is Good and evil does not exist. But that central religious verity happened to be a medical truth as well. For Eddy and her followers the understanding that all is Mind, not matter, and that neither sickness nor death is real represented the culmination of medicine's evolution and the harbinger of its future.

As many scholars have noted, such a spiritual understanding of reality proved to be Christian Science's most distinctive intellectual contribution to American religion. But Eddy's views of the mind-body relationship and her healing techniques owed much to the principles of homeopathic medicine and the practice of Maine mentalist Phineas Parkhurst Quimby. Before the appearance of *Science and Health* (1875), a textbook of her teachings, there is little evidence that Eddy thought of herself as more than a devoted disciple of Quimby as she adapted his teachings to new circumstances and defended them against charges of mesmerism or spiritualism—a Paul to his Jesus. Even after she completed the textbook, which probed beyond Quimby's decidedly practical tendencies, her message, despite the protests of her apologists, simply constituted a more Christianized version of Quimbyism. Such a recognition, however, should not diminish an appreciation of the magnitude of her accomplishments as formalizer of doctrine and founder of institutions, any more than the priority of Jesus' gospel detracted from Paul's decisive activities as theologian and missionary.

Eddy's ideas bore many similarities to those of a swarm of religious enthusiasts, mesmerists, spiritualists, and health reformers who, in the words of Christian Science historian Robert Peel, appealed to "a restless people reaching out blindly for new sources of power, of assurance, of spiritual hygiene."[14] Yet contrary to the dominant historical narrative, the nineteenth-century origins and early twentieth-century development of Christian Science in the United States were just as much entwined with the theories and practices of medicine as with Christianity. As a consequence, I seek to reposition our understanding of the movement's identity within American culture. Christian Science was a medicoreligious hybrid formed and sustained by its founder's and adherents' search for a physical well-being anchored in spiritual reality. But the American medical community's successful efforts to unify medicine under its banner of science pressured Christian Scientists to abandon much of their medical identity.[15]

Struggling to give a distinctive sound to her message, Eddy devoted much of her time after 1866 to teaching, healing, and writing; but her movement grew slowly, and when in 1879 she and her followers organized the Church of Christ, Scientist, in Lynn, Massachusetts, they numbered only twenty-seven. Undeterred, she moved her headquarters to cosmopolitan Boston, where she chartered the Massachusetts Metaphysical College (1881) to educate practitioners in Christian Science healing.

During the 1880s, mind healers of all kinds spread across the country from New England. Warren Felt Evans, whom Quimby had cured of a serious nervous disorder in 1863, devoted his life to the refinement, dissemination, and practice of mental healing, exerting through commentaries such as *The Mental Cure* (1869) and *The Primitive Mind Cure* (1885) a formative influence on the development of what came to be known as New Thought. Calling themselves by various names—mind curists, mental curists, and metaphysical healers—these adherents to New Thought diverged widely on the particulars of their views, but they shared a confidence in the power of the mind to cure disease and solve human problems and a belief in a quasi-religious, often idealist philosophy that seemed to explain the healing process. Eclectic and independent minded, these healers drew on the mystical writings of the eighteenth-century Swedish philosopher Emanuel Swedenborg, East Asian as well as Judeo-Christian teachings, and the marvels of spiritualism and science to supplement and defend their worldviews.[16]

Thousands of Americans discovered mind healing in Eddy's *Science and Health*. Eager for a cure for their diseases or the comfort of belief in a benevolent world, these persons often failed to notice any significant difference between Eddy and other mentalists, each with his or her own brand of doctrine. Concerned primarily with the broad picture and unworried by obscure ideological distinctions, large numbers of New Thought adherents and Christian Scientists drifted on a tide of ideas, changing their beliefs or practices to accommodate the latest book they had read or to emphasize a nuance they themselves had discovered.

Just as Eddy's overshadowing influence on Christian Science often reduced its history to biography, so scholars frequently have turned an analysis of the practice of Christian Science healing into an extended gloss on Eddy's prescriptive writings about healing. In contrast, here I describe what Christian Science practitioners reportedly did, not just what they were supposed to do, unveiling the many obscure faces and diverse sensibilities that helped make Christian Science healing a viable alternative to medicine. Ironically perhaps, such an approach also leads to a more nuanced appreciation of Eddy and what she, together with her followers, accomplished in Progressive Era America.

Many Americans first came to Christian Science seeking physical or spiritual health, and those who stayed found both a congenial theology and a practical Christianity as well as an avenue for professional advancement. As practitioners, these converts, mostly women, could look forward to lives of economic and professional security and a clear sense of purpose. Eddy and her writings provided the foundation for their beliefs and practices, but the hard knocks of experience often led them to adapt to the circumstances at hand. Christian Science practitioners inevitably fell short, but their efforts to make sense of their failures led them to try again. Through an examination of practitioners' correspondence and courtroom testimony, this book reveals what practitioners believed and practiced and what patients expected and experienced, and it exposes the dialectic implicit in the healing process. The often defensive nature of their court testimony presents its own biases, but the evidence garnered reveals a complex interaction between belief and practice that scholars rarely have noticed. Moreover, practitioners did not operate, as many Scientists would have one believe, oblivious to the medical world around them; they often offered patients religious healing framed in the forms and language of medicine. The business of healing, from training to day-to-day practice, furnished glimpses of the healing process as revealed in the relationship between practitioner and patient.

Here I expose the ways the practice of healing presented its own two-part variation on the theme of a Christian Scientist's life as trial. On one hand, a healing or "demonstration" furnished evidence to prove the innocence of reality and to convince patients of the unreality of the "false claims" of sin, sickness, and death. In this way the healings bore a similarity to homeopathic "provings," which supposedly demonstrated the power of minute medicinal doses. A healing for Scientists presented the supreme witness to the efficacy of their methods, whether judged by a practitioner, a patient, or the court of public opinion. On the other hand, if a healing apparently failed, it proved to be a trial as well. Practitioners found themselves confronted with evidence profoundly contradicting their view of reality and presenting to them the greatest challenge to their faith in Christian Science.

Busy with various Boston affairs and confronted by numerous local challenges, during the 1880s Eddy found it increasingly difficult to deal with her far-flung followers. Religious critics derided her "bogus" Christianity and, comparing her to the purveyor of a popular patent medicine for female complaints, dubbed her the "Lydia Pinkham of the Soul." Former students and eclectic mentalists altered her teachings and questioned her prophetic authority; Julius A. Dresser and his wife, Annetta Seabury, former patients and students of Quimby's, accused Eddy of stealing their teacher's ideas. Eddy, still convinced that God had revealed the truths contained in *Science and Health*, emphasized her Christian roots and asserted the unreality of matter more adamantly than ever. She also moved to unify her followers and define "orthodox" Christian Science by emphasizing her divinely sanctioned authority, clarifying doctrines, and solidifying organizations.

In an effort to provide regular guidance and encouragement to her distant students, in 1883 Eddy established and edited the *Journal of Christian Science* (renamed the *Christian Science Journal* in 1885), a monthly publication intended to shield her students from the "unorthodox" teachings of rival Scientists. She also encouraged the graduates of her Normal classes at the Metaphysical College to fan out across North America, teaching, healing, and establishing institutes of healing and instruction. In the key growth states of Iowa and Illinois alone, Eddy's students established sixteen institutes during the 1880s and 1890s. Occupying an intermediate level of church organization between Eddy and local churches until their closing before 1900, these institutes served as successful regional schools for the teaching and spread of Christian Science. To further guard her students from heresy, Eddy encouraged the formation of a National Christian Scientist Association in the spring of 1886. At annual meetings of the association, she exhorted her own students and those of other teachers to remain committed to her brand of Science.

Yet Eddy's admirers often chafed under what they believed to be her excessively authoritarian methods. Some, like A. J. Swarts, acknowledged her unique contributions to mind healing but insisted that individuals should search for truth outside *Science and Health* and the Bible and follow it wherever it led without continually seeking her approval. He suggested that mind healers unite Evans's Mental Cure and Eddy's Christian Science into Mental Science. Others, such as Emma Curtis Hopkins, Luther M. Marston, and Ursula N. Gestefeld (all former students of Eddy's), objected to Eddy's efforts to establish herself as the only true expositor of Christian Science and to her criticisms of their publications and their schools. Despite the fact that the teachings of these healers and their students differed in some respects from Eddy's, and despite the claims of many Christian Scientists to the contrary, these healers' shared emphasis on physical healings and their continued use of the name Christian Scientist, even after their separation or expulsion from Eddy's church, lead me to label them "generic" Christian Scientists.

The 1888 manslaughter trial of Boston practitioner Abby H. Corner, which involved the death of a mother and child during childbirth, rocked the fledgling movement, undermined Eddy's authority, and fragmented her followers into "true" and "pseudo" Scientists. I examine the circumstances surrounding that trial and its aftermath and present the first extensive analysis of the distinctive practices of metaphysical obstetrics—the ways Christian Scientists understood and managed childbirth. Eddy moved rapidly to insulate herself from the unfortunate Corner affair, to rethink and ultimately reject the specialty of metaphysical obstetrics, and to reorganize the institutions that projected her message into the world. Although as a result her movement fragmented into orthodox and heterodox Christian Scientists, the cadre that remained loyal provided the solid core necessary for future advancement. Nonetheless, the post-Corner questions among orthodox Christian Scientists about Eddy's authoritarian and arbitrary leadership led to a major schism that shattered the church. In response Eddy disbanded all major church organizations and retired to New Hampshire for a period of spiritual reflection.

From her retreat near Concord, Eddy embarked on an intensive campaign to stimulate the growth of her movement and to place it on a sounder organizational footing by routinizing her authority among the orthodox. In 1892 Eddy established a central church organization, the Mother Church, in Boston and appointed a board of directors to oversee its affairs and implement her instructions to organize evangelistic lectures and monitor educational standards. And in 1898 she founded the Christian Science Publishing Society to spearhead worldwide evangelism through the printed word, including the *Christian Science Monitor* (1908). These evangelistic activities enhanced the growth of Christian Science, spreading it across the oceans from America's urban areas to touch Europe and Asia by the early twentieth century.

It was one thing for Eddy and her followers to doubt themselves or to be attacked by enemies; it was quite another to find family and trusted friends turning against them. Therefore, perhaps more profoundly than other trials, the incessant bickering, rivalries, and schisms that dogged the history of Christian Science continually challenged the movement's future. Repeatedly, such conflicts decimated the church's membership and threatened to bring an end to its mission. But phoenix-like, Scientists survived, and after reflection and careful reassessment of beliefs and practices, they re-created themselves to emerge again into American public life.

The official tenets of orthodox Christian Science, settled in the midst of these public disputes, subsequently underwent little change. In the eighty-ninth and final edition of the *Church Manual* they appeared as follows:

1. As adherents of Truth, we take the inspired Word of the Bible as our sufficient guide to eternal life.

2. We acknowledge and adore one supreme and infinite God. We acknowledge His Son, one Christ; the Holy Ghost or divine Comforter; and man in God's image and likeness.

3. We acknowledge God's forgiveness of sin in the destruction of sin and the spiritual understanding that casts out evil as unreal. But the belief in sin is punished so long as the belief lasts.

4. We acknowledge Jesus' atonement as the evidence of divine, efficacious Love, unfolding man's unity with God through Christ Jesus the Wayshower; and we acknowledge that man is saved through Christ, through Truth, Life, and Love as demonstrated by the Galilean Prophet in healing the sick and overcoming sin and death.

5. We acknowledge that the crucifixion of Jesus and his resurrection served to uplift faith to understand eternal Life, even the allness of Soul, Spirit, and the nothingness of matter.

6. And we solemnly promise to watch, and pray for that Mind to be in us which was also in Christ Jesus; to do unto others as we would have them do unto us; and to be merciful, just, and pure.[17]

Although much of the language used here was traditionally Christian, its meaning often varied from that of historical Christianity and always needed to be interpreted through a study of *Science and Health* and *Key to the Scriptures* (1883).

Today Christian Science Reading Rooms still provide, amid the bustle of hectic urban life, a quiet place for reading, reflection, and spiritual healing. Christian Science churches continue to hold eleven o'clock services on Sunday, not unlike those in other Protestant denominations, with singing, praying, and reading from the Bible and *Science and Health*. On Wednesday evenings the believers gather for services similar to Baptist prayer meetings to encourage fellow believers and to strengthen their own experience by testifying to what Christian Science has done for their lives. Most Scientists still consult only practitioners when ill, and although some children still die while under a practitioner's care, an unsteady truce prevails between American medicine and Christian Science. Americans have acquiesced to a tenuous legal standing for Christian Science healing, but the ambivalence of many Americans about this type of religious healing persists. The lot of Christian Scientists appears to remain one of lives on trial.

PART 1

*The World of
Christian Science Healers*

Mary Baker Eddy

Patient, Healer, Teacher

Any person desiring to learn how to heal the sick can receive of the under-
signed instruction that will enable them to commence healing on a principle
of science *with a success far beyond any of the present modes. No medicine,*
electricity, physiology or hygiene required for unparalleled success in the
most difficult cases. No pay is required unless this skill is obtained. Address,
MRS. MARY B. GLOVER, *Amesbury, Mass., Box 61.*
—*Banner of Light,* 1868

In April 1878 Mary Baker Eddy, short of money and jealous for the
reputation of Christian Science, filed suit against two former students,
George H. Tuttle and Charles S. Stanley, for failure to fulfill the terms of
an agreement they had signed with her in 1870. Rather than devoting her
own time to healing, since 1868 Eddy had advertised healing classes in
Moral Science (early Christian Science), which she claimed "was efficacious beyond all other known medical science or practise, and in effect
infallible when rightly understood and applied." She had required all six
students in her first Lynn, Massachusetts, class to sign a statement similar to the following: "We the undersigned, do hereby agree in consideration of instruction and manuscripts received from Mrs. Mary B. Glover
to pay her one hundred dollars in advance and ten per cent. annually on
the income that we receive from practising or teaching it. We also do
hereby agree to pay her one thousand dollars in case we do not practice
nor teach the above named science that she has taught us." In her bill of
complaint, Eddy claimed that she had fulfilled her obligations, but once
Tuttle and Stanley had "jointly practised and taught said method of

healing and said science," they had refused, despite her repeated requests, to fulfill the contract. She asked the court to examine the defendants' account books to determine what they owed her, and then to require payment with interest.

Stanley challenged certain specifics of the suit with the result that Eddy filed an amended bill on 1 January 1879 in which she conceded that she did not know for a fact that the defendants had practiced what she had taught them. After another exchange of filings, the two parties agreed to allow George F. Choate, a court-appointed referee, to decide the case. Although the events in question had occurred nine years before, in an effort to get at the truth Choate took testimony from Stanley, Tuttle, Eddy, and her former healing partner Richard Kennedy.

Choate discovered that Tuttle had paid one hundred dollars in advance for his twelve healing lessons, but Eddy had allowed Stanley to pay twenty-five dollars a week on an installment plan.[1] After Stanley had received several lessons and made his second payment, he began to express "some doubt or distrust of her [Eddy] or her science," whereupon Eddy stopped teaching Stanley and took back the manuscripts she had given him. Recalling the incident somewhat differently, Eddy reported that she had not dismissed Stanley; rather, he had left before completing the lessons because he thought he already knew enough "to go right into practice."

Subsequently, Stanley practiced healing by various means and at times referred to himself as a "Christian Scientist." However, Choate found that Stanley had not used any of Eddy's distinctive teachings or practices "except certain manipulations which she and others have at times used, but which she discards at the present time and claims are no part [of] her system or science." Besides, said the referee, "I do not find any instructions given by her [Eddy], nor any explanations of her 'science' or 'method of healing' which appear intelligible to ordinary comprehension" or which, in his opinion, could have been of any value in preparing Stanley to heal the sick. Choate concluded that Eddy should not be awarded anything, and the court so ordered. Eddy immediately appealed the decision, but after having second thoughts about the propriety of pursuing the case, she withdrew the appeal. The court then entered judgment for the defendant Stanley on 9 February 1881, requiring Eddy to pay his court costs.[2]

The suit against Tuttle and Stanley was just one of several involving

Eddy and her supposed enemies during 1878. She similarly charged former students Richard Kennedy and Daniel H. Spofford; she supported Lucretia L. S. Brown's charges that Spofford had maliciously attacked her mind with mesmerism; and she became entangled in the state's indictment of her husband, Asa Gilbert Eddy, and her student Edward J. Arens on two counts of conspiracy to murder Spofford.[3] Unlike the court trials that would begin ten years later in which the "world" would plague Christian Scientists, in these cases the proto-Scientist Eddy and her followers attacked their enemies. But like those later trials, which reveal much about the beliefs and practices of Christian Science practitioners, these trials have much to tell about Eddy's early healing and teaching career. Despite Eddy's reputation as a powerful and authoritative leader, her actions often hid insecurities rooted in years of personal struggle with illness, poverty, and professional rivals. In efforts to defend what she saw as her threatened interests, she often used the courts. As we trace the material and spiritual roots that allowed Eddy to give birth to Christian Science, survey the early evolution of her thinking about religious healing, and examine the development of her career as healer and teacher, we shall see that she often aggressively denied her intellectual dependence on Quimbyism, asserted her priority in discovering Science, and in countless other ways struggled to create and preserve her public identity.

A Spiritual and Medical Odyssey

Born near Concord in the township of Bow, New Hampshire, on 16 July 1821 and raised in Sanbornton Bridge, only about eighty miles from Portland, Maine, Mary Baker Eddy experienced many of the religious tensions and medical vagaries of Jacksonian America.[4] (Married three times, Eddy went by numerous names over her lifetime: Mary Morse Baker, Mary Baker Glover, Mary M. Patterson, Mary Baker Eddy. To avoid confusion, I refer to her as Mary when discussing her childhood and youth and as Eddy during her adult years.) As a child and young woman, she suffered from a persistent host of physical and mental illnesses and religious anxieties that at times severely restricted her activities, dampened her energetic spirit, and hindered her formal education. However, she also appeared on various occasions throughout her life to use illness for social advantage and to gain the space necessary for creative thought.[5] At one time or another she endured colds, fevers, chronic dyspepsia, lung and liver ailments, backache, "nervousness," gastric attacks, and "depression" that propelled her on a lifelong search for a remedy for disease. An ambitious child who hoped one day to make her mark as an author, she read and studied when possible, but her ill health and weak constitution prevented her from receiving more than sporadic formal education and occasional tutoring from family members.

Disappointed by the ineffective treatment she received from local physicians, Mary tried self-help remedies and experimented with sectarian cures. During the 1830s, when the health reformer Sylvester Graham was touting coarse whole wheat bread, pure water, and a vegetarian diet as both remedy and prevention for disease, Mary tried Graham's cure; but although she followed it fairly consistently over the next twenty-five years, it never gave her lasting relief.

Many of the experiences that early fed Mary's religious identity grew out of a similar tendency to question or experiment with received theology. When she was about eight, she heard unknown voices calling her name. On the advice of her mother, Mary responded in the words of Samuel: "Speak, Lord; for thy servant heareth" (1 Sam. 3:9).[6] Mary's mother, who did not believe that prophets had ceased to appear after the apostolic age, thought her child might represent a fulfillment of the "last days" prophecy of Acts 2:17: "Your sons and your daughters shall prophesy, and your young men shall see visions." Such advice encouraged Mary to infuse spiritual meaning into the experiences of everyday life, and following her mother's advice, she often took her mental and physical torments to God in prayer. Doubts about the justice of divine predestination led Mary into painful debates with her father, deep mental anxiety, and spells of sickness severe enough to require a physician's attention. Though at age seventeen she refused to admit that a merciful God would damn some souls to eternal hell, the pastor of her parents' Congregationalist church still admitted her into membership.[7] This strong desire for belief in a just God undoubtedly contributed to her later conception of a universe without evil.

Eddy's biographers have made much of the contrasting religious emphases of Mary's parents. Although both parents were long-standing Congregationalists, in Mary's memories their theological orientations differed considerably. Abigail rooted her religious faith in a loving God who eagerly sought to relieve the trials of his children, while her husband, Mark, measured true belief by adherence to theological tenets and strict moral behavior. Whether Mark and Abigail held views as divergent as Mary recalled will probably remain hidden. What is apparent, however, is that much later, while adapting her beliefs to form a new religious movement, Mary blended these two strains of Christianity into a balanced mixture of divine grace and justice.

Aside from the fact that Mary may have been more sickly than the typical antebellum youth, little about her formative years points to the career that lay ahead. Her literary aspirations and hesitant poetry, her spotty formal education, and the intense religiosity she infused into even mundane experiences were typical of many contemporary New England girls. Only with hindsight might one see the potential connection between sickness and Christianity that became the focus of Christian Science.

New England, rich in tradition but also fertile for the political, religious, and medical innovations that animated nineteenth-century America, nurtured and then gave flight to Mary Baker Eddy. Jacksonian democrats, religious revivalists, and medical sectarians all damned the authority of "experts" and praised the principles of individualism that characterized the "age of the common man." Yet despite their occasional criticism of scientists, churchmen, and physicians, these nineteenth-century Americans were fascinated by the wonders of the spiritual and natural worlds. Guided by their commonsense belief that God, the author of both nature and revelation, would never contradict himself, they often reconciled ingenious scientific discoveries with dusty religious doctrines by revising spiritual truths or devising new spiritual pathways. Questions about human nature and the care of the body and soul preoccupied many nineteenth-century Americans and led to repeated attempts to reconcile science, medicine, and religion. Each decade seemed to bring some new understanding of human nature that required adaptation to Christianity: phrenology in the 1830s, mesmerism in the 1840s, and spiritualism in the 1850s.[8]

During the late eighteenth century the German physician Franz Joseph Gall, as a result of his anatomical observations and conjectures, concluded that the brain was composed of several different organs, each responsible for a different mental capacity or moral characteristic. Based on this localization of brain function, Gall and his fellow phrenologists believed that examining the external surface of the skull would expose bumps that revealed which brain organs were robust and which were frail. Such ideas held dangerous determinist implications, but then Johann

Gaspar Spurzheim and George Combe, two of Gall's followers, transformed phrenology into a guide for character development. Through exercise, they claimed, one could strengthen the mental organs responsible for positive characteristics and weaken the negative influences. When Spurzheim and Combe lectured in the United States during the 1830s, Americans received the new science with great enthusiasm as further evidence of God's great design.

Mesmerism, or animal magnetism as it was popularly called, proved specially effective in helping Americans relax nervous tensions, relieve spiritual ennui, and restore physical health. First widely publicized across New England by the public lecture tour of Charles Poyen in 1836, mesmerism (similar to the hypnotism of today) fascinated Americans with its bizarre manifestations of clairvoyance, somnambulism, anesthesia, and ecstatic utterance. By touching their patients or simply through mental suggestion, mesmerists could diagnose disease and often effect a cure. Professor and peasant alike were attracted to the phrenologists, magnetic healers, spiritualists, and preachers who used mesmerism to awe, cure, or hoodwink.

By the 1850s the spirit rapping of the Fox sisters had mushroomed into arguably the fastest-growing popular religion in America. In a commonsense way, not unlike phrenologists and mesmerists, spiritualists believed in a reality that transcended the mundane world of human sense. Mental gymnastics, mesmeric fluids, or the disembodied spirits of the dead each lifted Americans to consider the grandeur of God and the limitations of human knowledge. But the veracity of each could be tested by the simple, empirical tests available to common men and women. Trance mediums and healing mediums abounded, each serving as a conduit to the spirit world and the untold wisdom of the ages. Claiming to diagnose disease and provide prescriptions from the beyond, many spirit mediums set up healing practices and earned comfortable incomes.

The unexpected death of Eddy's first husband, George Glover, just six months after their marriage in December 1843, and the subsequent birth of their son two and a half months later thrust upon Mary responsibilities for which she was ill prepared and triggered months of sickness and melancholy that often reduced her to a bedridden invalid. A phrenologist read her skull in 1848, and she learned that she possessed strength in *"Philosophy*—truth combined with conscience—and affection," but the good news only temporarily buoyed her spirits.[9] Still searching for relief, she dabbled in mesmerism and became familiar with its apparent influence over human thoughts and behavior and its power to cure illness, and she received homeopathic treatments.

By the 1840s homeopathy had acquired a stable number of adherents since its appearance on American shores in 1825. Samuel Christian Hahnemann, a German physician of the late eighteenth century, founded homeopathy in response to the overmedication of his day. He carefully conducted controlled experiments on drugs (later called "drug-provings") to observe and record their effect on a patient's symptoms. From these studies Hahnemann concluded that a patient could be cured when administered a drug that produced the same symptoms as the disease (the doctrine of *similia, similibus, curantur*). Subsequently he added the law of infinitesimals to his system. According to this principle, the more diluted a drug, the more potent its effect on the disease. He coined the term "allopathy" to refer to orthodox physicians who, in his view, pursued an "other" theory of drugging.[10] Eddy adopted the term and used it throughout her life to identify the system of medicine embraced by regular physicians.

The 1840s, which had begun with the death of her beloved brother Albert in 1841, ended as tragically with the death of her mother in November 1849, and the 1850s brought more heartache and misfortune. Feeling unwelcome in her father's home after his remarriage and unable or unwilling to care for her willful and often disobedient son, Eddy turned George Jr. over to the care of another family in May 1851. The separation of mother and child precipitated a severe decline in her health that lasted a year. Her engagement and subsequent marriage in 1853 to a struggling dentist named Daniel Patterson seemed to raise her spirits and strengthen her health, but when her new husband made it clear that he did not want a stepson in his house, her incapacitating ill health returned and she remained practically housebound as Patterson pursued his itinerant trade. For most of Eddy's life ties with her son remained virtually nonexistent, and although she heard from him in 1861, she did not see him again until 1879. Her ill health, which confined her to bed for a year and a half beginning in 1857, persisted throughout the decade. She turned again to homeopathy and vowed to God that if he healed her she would dedicate herself to helping others. Throughout the decade she studied the homeopathic doctrines of *similia* and *minima* in *Jahr's New Manual of Homeopathic Practice*, tested a variety of medicinal dosages, and treated herself, friends, and acquaintances with homeopathic remedies. When her patients recovered despite taking solutions so diluted that they contained virtually no medicine, she concluded that their faith in medicine—not the medicine itself—had caused their recovery (or so she claimed later in life).

Christian Science holds abundant evidence that Eddy's spiritual sensibilities and her early forays into the medicine of her day contributed much to the formation of her thinking. That no healing system seemed to give her a permanent cure reinforced her distrust of the standard medical options and encouraged her to tinker with them or to try out the latest fads. Her experiences with homeopathy led her to doubt the efficacy of drugs and to suspect the power of mind on body, suspicions that her encounters with mesmerism seemed to confirm. Steeped in the Victorian Christianity of her day but doubtful about some of its emphases, Eddy came to believe that God might have called her to reform medicine and to restore the gospel of a loving God to its rightful place among Christian doctrines.

A Quimby Disciple

By 1861 Eddy still had not found a long-lasting cure for her maladies, and she seemed to be running out of options. Despite her growing desperation, however, her continued use of Graham's bland vegetarian diet and her more recent experiences with mesmerism and homeopathy encouraged her to consider two final possibilities—hydropathy and Quimbyism. Hydropaths believed that the human body inherently possessed the power to prevent and cure disease. When ill, however, the body's self-curative powers could be vitalized by copious amounts of water, taken internally and applied externally, a nonstimulating diet, sunlight, exercise, and relaxation. During the 1850s Graham's health reformers had forged links with hydropathists, and Eddy, apparently believing that her dietary regimen had staved off the most serious consequences of her "spinal inflammation and its train of sufferings—gastric and bilious," checked into Vail's Hydropathic Institute in Hill, New Hampshire, in June 1862.[11] She stayed nearly three months, but her condition deteriorated until she could barely walk. Just then a former Vail patient revisited the hydropathic institute to sing the praises of Phineas Parkhurst Quimby (1802–66), a famous mental healer in Portland, Maine, who had studied and experimented with magnetic healing since 1838. Eddy heard the glowing reports and decided to try Quimbyism.

Even before deciding to enter Vail's Institute, Eddy had solicited treatment from Quimby, but when he failed to respond to her letter, she had opted for douches and sitz baths. She now concluded that she should have pursued Quimby, and by the end of the year she had arrived in Portland and placed herself in his care. Convinced that a patient's trust in the healer effected cures, Quimby attempted to establish a rapport with his patients by experiencing their symptoms, massaging their heads or limbs, and speaking encouraging words. Although he practiced healing for less than two decades, he exerted an important influence on the birth of both spiritual healing and mind cure in America. During 1862 and 1863, three major contributors to the early development of American mind healing—Eddy, Warren Felt Evans (1817–89), and Julius A. Dresser (1838–93)—received treatment at his hands.[12] Eddy underwent an immediate, though only temporary, cure with Quimby's help, and after returning to Maine for further treatments and lessons in mental healing during the winter of 1863–64, she departed to exercise her newfound talents.[13]

During her early practice of Quimbyism, not only did Eddy suffer recurrences of her own maladies, she annoyingly acquired her patients' symptoms vicariously. For example, even though she failed to cure her nephew Albert of his tobacco habit, she acquired a desire to smoke. Early in 1864, however, when she healed Mary Ann Jarvis of an unidentified condition and did not acquire her symptoms, Eddy was so elated that Jarvis's healing signaled for Eddy the dawn of Christian Science, much as her own healing would do after 1866. Further healing successes followed through the end of 1865, but her Quimbyist methods proved far from perfect. Most troubling, she could not heal herself. Whenever she remained in Quimby's care she seemed to recover, but when she returned home she relapsed.[14] Then in January 1866 Quimby died.

Less than a month later, Eddy fell on the icy streets of Lynn, Massachusetts, and was knocked unconscious. Her friends feared the worst when the ministrations of a homeopathic physician did little to alleviate her severe head, neck, and back pains. Hopeless and depressed, Eddy turned to the Bible for encouragement. While reading the Gospel account of Jesus' healings one day, she discovered the "healing Truth" of Christian Science and experienced a spontaneous recovery. Later, through the eyes of faith and refashioned memories, she testified: "Ever after [I] was in better health than I had before enjoyed."[15] As the years passed, Eddy turned this recovery into a "Damascus road" experience that provided a sign of God's recognition and strengthened her mother's claim that she had been singled out by God for a special mission. But the nature of this "religious healing" also led her to discern a purpose behind her aimless wanderings down nineteenth-century medical and religious byways. Albeit tentatively at first, she rethought her early experiences with homeopathy, mesmerism, Quimbyism, and Christianity in the light of her recent spiritual "discovery" and, with a Bible at her side, began to outline the views of God, sin, sickness, and healing that would later constitute her new religion of health.[16]

Eddy's spiritual discovery did not alleviate her persistent difficulties with life, however. On 15 February 1866, two weeks after her fall in Lynn, beset again by spinal and intestinal problems, she wrote to Julius Dresser, a fellow student of Quimby's, beseeching him to take her as a patient. Dresser replied with words of encouragement and an injunction not to trust in "matter," but he refused to treat her.[17] To make matters worse, her ne'er-do-well second husband, Daniel Patterson, from whom she finally separated in August, left her on her own with no money to pay the rent. Thus began a four-year period during which Eddy, forced to depend on herself and God, worked to fashion out of her healing experiences a career that might ensure her economic survival and fulfill her spiritual calling.[18]

She found lodging in the boardinghouses or homes of compassionate spiritualists and those "interested in the practical business of healing."[19] When their charity wore thin, she traded her teachings—which at this point differed little from what she had learned from Quimby—for room and board. In this way her first student, Hiram Crafts, established a short-lived practice in Taunton, Massachusetts, but Eddy needed a more dependable way to attract students. During the summer of 1868, therefore, she advertised as a teacher in the spiritualist journal *Banner of Light*, confidently claiming for her healing method "a success far beyond any of the present modes" and an "unparalleled success in the most difficult cases."[20] Even then only a couple of students responded. Despite her precarious health and finances, she managed to complete *The Science of Man, by Which the Sick Are Healed, or Questions and Answers in Moral Science*, a pamphlet written in a question-and-answer format that she used to instruct six students during fall 1870 in Lynn, Massachusetts.

The Science of Man contained the earliest published form of her healing system, which at the time Eddy called "Moral Science." (She later called her teachings "Metaphysical Science" or simply "Metaphysics" before finally settling on "Christian Science.") The purpose of Eddy's instruction at this time was to turn her students into healers, and to do that they needed to understand the nature of reality. According to Eddy, God is not a person but a principle of "wisdom, Love, and Truth" possessing life and intelligence, and a human is a "reflex shadow" of that principle. The central point of Science was that "man has no substance or intelligence, that these belong to the Soul [God], and the Soul reflects itself in man." These truths, however, do not become an experienced reality for an individual until they are acknowledged. Until that time, therefore, even for a materialist, what a human believes to be so constructs reality. Hence she stressed the importance of believing truer and better things until belief is replaced by knowledge, which is God. Sickness disappears when one acknowledges it as a belief that must be replaced by knowledge—not so much a knowledge of a healthy body as a knowledge that all is God and a reflection of him. Healing is the grasping of that understanding.[21] Often these instructions seemed extremely abstract, and to help both healer and patient understand that the healer was effecting a cure, clairvoyant healers, mesmeric healers, rubbing mediums, Quimbyists, and early students of Eddy's massaged or touched the offending area of a patient's body. Eddy encouraged her students to understand that "rubbing has no virtue[;] only as we believe and others believe we get nearer to them by contact and now you would rub out a belief, and this belief is located in the brain." Nonetheless, "as an 'M.D.' lays a poultice where the pain is" so the healer should lay her "hands where the belief is to rub it out forever."[22]

The Kennedy Partnership

Among the students Eddy instructed from mid-1868 to early 1870 was Richard Kennedy, a young and energetic disciple who early caught Eddy's eye and furnished a new way for her to earn an income and spread her healing system. After completing his coursework, Kennedy agreed in February "to pay Mary M. B. Glover one thousand dollars in quarterly installments of fifty dollars commencing from this date."[23] Another recent student, Sarah Bagley, signed a similar contract "to pay Mrs. Glover 25 per cent of all the money she earned by healing, in return for the instruction she had received," but Eddy planned to gain more from her Kennedy partnership than just a dependable income.[24] When the two moved into offices in Lynn, Massachusetts, in May 1870, the steady flow of satisfied patients soon created a pool of willing students for the three classes Eddy held in Lynn. She charged the members of her first class, which included Tuttle, Stanley, and four others, $100 each and a percentage of the income from their future healing practices, but she charged a flat fee of $300 to the seven students in the last two classes, each of which included twelve lessons over three weeks. With Kennedy's practice thriving and her share of the first year's receipts totaling $1,742 after deductions for rent and all business and living expenses, she could even discount the tuition for some of her students. Profits continued, and by the end of the second year in Lynn, Eddy had $6,000 in the bank.[25] Financially, the partnership was a ringing success; in other ways it proved a disappointment.

In 1862 Eddy had written a letter to a Portland newspaper defending her mentor, Quimby, against some critics who maintained he practiced nothing more than spiritualism and others who claimed he was a mesmerist.[26] Since then, as Eddy had proselytized for Quimbyism, she had associated with many spiritualists herself, and now similar charges resurfaced in Lynn, as "many of her friends, and some of her pupils, were inclined to believe" she was a spirit medium.[27] In her effort to distinguish healing mediums from "scientific" healers, she prepared for her students a paper that expanded her earlier Portland remarks. Many spirit mediums, she believed, were frauds, but those who operated based on true clairvoyance effected cures because their patients believed the medium could "see" the material cause of their disease and cure it. If one believed in something hard enough, one could make it so. In contrast, she continued, "when I examine the sick I go upon the Principle of Science, that sickness is not truth, that it has no locality, and is nothing. Mark! The reverse of their views!" But the danger for harm in the clairvoyant situation was great, for false beliefs can bring evil instead of good. What, Eddy's readers must have wondered, preserved one's beliefs from doing harm? She replied:

> I cannot use this Science to do evil with it. The moment I should attempt this error I should lose Science, inasmuch as I would be working in error and not in Truth. Hence, the impossibility for me to give a demonstration in Science that works ill to myself or neighbor, as also the impossibility for

me to visit the absent in a speculative mood, or out of curiosity, or a desire to influence a mind to any evil thoughts or actions. Why I cannot do this is because I cannot act or think in matter where sin is or the belief called sin, which is error, if I am acting or thinking in Science. Now, I cannot give the phenomena of belief when I give the phenomena, above alluded to, but I give these manifestations in Science, and, whereas, I once gave them in belief, or matter, I now give them understandingly; whereas, I once gave them darkly, not in knowing how I did it, I now give them in light.[28]

In Eddy's earlier defense of Quimby she had stated that she understood his healing principle "dimly," as "trees walking" and "just in proportion to my right perception of truth is my recovery."[29] She now seemed to be saying that in her 1866 recovery she had discovered the healing principle to be a self-authenticating knowledge that bypasses the need for human belief. But rather than the more typical self-authentication of a mystical, religious encounter with a holy "other," the authenticity of Science came through the demonstration of its truth in healed bodies and changed lives. Because Eddy knew that evil does not exist, it had become impossible for her to do evil; because she knew that only truth exists, it was impossible for her to teach error.

It is doubtful that this explanation convinced many skeptics, but during 1872 Eddy experienced a second and more public challenge to her fragile authority when a former student, Wallace W. Wright, accused her in a Lynn newspaper of practicing mesmerism and her associate Kennedy refused her command to reduce the amount of rubbing and manipulating in his practice. Stung by Wright's charge and Kennedy's insubordination, Eddy dissolved her partnership with Kennedy, ceased manipulation, and branded as mesmerism all schools of metaphysical healing that differed from her own. By the time she taught her next healing class, in 1875, she had revised the 1870 edition of *The Science of Man* to exclude all reference to manipulation, but when she finally published the pamphlet in 1876 and released it to a wider public, she included a section that discussed her Kennedy experience and labeled him and others like him malpractitioners for continuing manipulation.[30]

Eddy believed, as did some of her contemporaries, that mesmerists could marshal the magnetic forces of animal magnetism for either good or evil. Increasingly embattled and suspicious, she gathered her most faithful followers around her to ward off the affects of "malicious animal magnetism" that she felt her enemies were directing against her mind. During these trying years, she gained great courage and support from her future husband, Asa Gilbert Eddy, a converted sewing-machine salesman, whom she married in 1877.

The Creation of a Textbook

While it lasted, Eddy's partnership with Richard Kennedy supported her teaching and allowed her to accumulate a nest egg that freed her for the study and reflection that culminated in the publication of *Science and Health* (1875), the textbook of Christian Science.[31] From May 1872 to October 1875 Eddy industriously wrote and rewrote drafts of *Science and Health*, based on what she claimed to be divine insights gained through study of the Bible and early morning visions, which usually occurred at times of severe stress and anxiety and provided supernatural answers to problems encountered in her work. Adamantly rejecting the claims of critics who found similarities between her writings and those of mental healers or idealist philosophers, Eddy asserted that she had "consulted no other authors and read no other book but the Bible for three years," adding that "it was not myself, but the power of Truth and love, infinitely above me, which dictated 'Science and Health with Key to the Scriptures.'"[32] Using an allegorical hermeneutic that she generously applied to the Bible, she interpreted her visions and dreams in terms of her developing doctrines and metaphysical principles.[33]

Embracing a radical idealism, Eddy affirmed that there is "no Life, Substance, or Intelligence in matter. That all is mind and there is no matter."[34] Humans and the physical universe are really perfect ideas that emanate from God and reflect his harmonious and eternal existence. Only God, his manifestations, and the synonyms that express the completeness of his nature—Mind, Spirit, Soul, Principle, Life, Truth, and Love—exist; all else, especially body, matter, error, and evil, is merely illusion, the nonexistence of which is proven as humans grow to reflect God. Eddy believed that the first-century Christians understood these truths and used them to defeat sickness, error, and death; the recent reappearance of these truths in the teachings of Christian Science signaled impending doom for all contemporary evil.[35]

By calling her teachings Christian Science, Eddy invoked two widely influential nineteenth-century ideologies. She believed that Christianity could be revitalized by her discovery of the truths that had allowed Christ to heal the sick and raise the dead in New Testament times, and she appealed to the methods of science to prove the truth of her claims through reason and the empirical evidence of healed bodies. She claimed that a kind of deductive logic unified her teachings into a convincing system of doctrine. For example, if God is all that exists and he is spirit, then matter, sickness, mental illness, and death do not exist. If God is all that exists and he is good, then evil and sin do not exist; claims for their existence merely reflect the tenacity of false beliefs and the undue attention paid to the false reports of the senses. Eddy asserted that it became easier to grasp the authenticity of such claims when one observed the concrete results of a healed body or a transformed nature. Calling such evidence a "demonstration," she concluded that "the best sermon ever preached is Truth demonstrated on the body, whereby sickness is healed and sin destroyed."[36]

At least as early as the first edition of *Science and Health* (1875), Mary Baker Eddy staked out her position against both the organized Christianity and the medicine of her day. Both clergy and physicians, she believed, had lost sight of the truths revealed by Jesus Christ and hence struggled blindly and often ineffectively against sin and sickness. Eddy's attitudes toward nineteenth-century physicians verged on outright derision, for example, when she claimed that "when there were fewer doctors and less thought bestowed on sanitary subjects there were better constitutions and less disease."[37] In addition to denouncing the drugging of allopaths, the bathing of hydropaths, and the eating habits of the Grahamites, she saved a special dose of invective for all mentalists and mind curists, whom she regarded as at best frauds and at worst criminals. Homeopathy, she believed, stood midway between the darkness of allopathy and the radiance of Christian Science, but since the truth of Science had dawned, homeopathy too must be put aside. At times she acknowledged that many physicians possessed "great philanthropy of purpose," but she urged them to "make their endeavors more effectual by changing their basis of action" to the metaphysics of Christian Science.[38] Although her attitude toward these false teachers mellowed as time passed, she never came to see them as allies in the human struggle against evil.

Christian Science, with its philosophical denial of matter and evil, almost openly invited some adherents to embrace human immorality, and such would have been the case more often had Eddy not retained the strict principles and practical exercises of her father's fierce moralism.[39] Eddy believed that her scientific reform of Christianity constituted a restoration of the full apostolic faith, which had fallen into decay and abuse over the past centuries, but adherents required a spiritual regimen of regular daily study of the Bible and *Science and Health*, prayer, and the prod of suffering to unlock their Scientific understanding and demonstrate the reality of health. Although suffering does not appease God and ideally speaking does not exist, it inevitably follows sin, sickness, and death, jolts one from one's illusions about reality, and moves one down the path of spiritual understanding. She insisted on strict moral uprightness, even in matters of diet. Echoing Graham, she denounced "depraved appetites for alcoholic drinks, tobacco, tea, coffee, [and] opium."[40] This strong emphasis on personal morality and exclusive commitment to the Bible and *Science and Health* later distinguished Eddy and her followers from many generic Christian Scientists and mind healers.

In April 1875 Eddy emerged from her self-imposed isolation to teach her first class in four years and to present her first public lecture, "Christ Healing the Sick," followed by weekly Sunday lectures through the month of June. Impressed by her presentations, a small group of supporters, agreeing to call themselves Christian Scientists, pledged $10 a week for one year "for no other purpose or purposes than the maintenance of said Mary Baker Glover as teacher or instructor" and "the renting of a suitable hall and other incidental expenses."[41]

With her first "organization" supporting her, Eddy began to search for the funds necessary to publish her 456-page manuscript, *Science and Health*. Passing the hat among her followers, by the end of the summer she had acquired the $2,285.35 needed in gifts and loans, and the first edition appeared in October 1875. However, despite extensive advertising through circulars and newspapers, discounting the price to $1, and distributing free copies to the religious press, it took over a year and a half to dispose of the one thousand books.[42] Though disappointed by the less than excited press reviews and the book's slow sale, Eddy and her followers remained heartened by their knowledge that Truth rarely finds an immediate welcome.

Christian Science Teacher and Preacher

The second half of the 1870s proved very trying for Eddy, as disenchanted followers pointed to inconsistencies between her words and deeds and deserted her for other religious healers. As they had in the past, Asa Gilbert Eddy and a handful of followers remained faithful and joined her struggle against the forces of "malicious animal magnetism" that they believed rival metaphysical healers directed toward them.

Eddy again took up teaching, formally instructing about thirty-five students between 1876 and 1881, while Asa guided about a dozen more in 1878. From these students and what remained of Eddy's earlier converts, there emerged a Christian Scientist Association. Though it did not hold public services, its constitution established the first formal organization of the movement. Eddy's income remained limited, however, and in a series of ill-fated attempts to retrieve lost income, she filed suits against former students who had broken their contracts to share their income with her. As we have already seen, these efforts did little to remedy her financial straits, but they did much to spread ill will and fuel rumors about Eddy's acquisitive nature.

Despite these diversions, Eddy delivered a series of Sunday afternoon sermons at the Baptist Tabernacle in Boston during the winter of 1878–79. Encouraged by the growing public interest in Christian Science, in April 1879 she and twenty-six association members took the decisive step of establishing a church, later designated the Church of Christ (Scientist), with Eddy as pastor. Lynn remained the hub of Eddy's activities, but not for long. Feeling the tug of cosmopolitan Boston, she and her followers worked to establish their reputations there by starting healing practices, distributing Christian Science literature from door to door, and holding publicly advertised lectures.

In her teachings Eddy denied the reality of the material realm, argued that health meant the complete "harmony of man," and urged her listeners to experience victory over sin as well as relief from physical maladies. Ironically, however, her own cure and moderate financial success provided a precedent for the movement's emphasis on the physical advantages of Christian Science healing. Initially at least, most of Eddy's early followers did not seek spiritual healing so much as physical healing

of deformity, infectious disease, or chronic discomfort. When such healings occurred, they were her best advertisement, and her income grew
with the spread of Christian Science teachings.[43] To develop the potential
interest inherent in the practical advantages of Christian Science healing,
Eddy obtained a charter for an institution of Scientific healing, the Massachusetts Metaphysical College. With its opening in Boston in May
1882, Eddy and her followers had stepped onto a new stage and opened a
new act in the production of Christian Science.

Troubled in body and soul, Mary Baker Eddy fulfilled her spiritual
calling and ensured her economic survival by adapting Quimbyism into a
religion of health and well-being. As the years passed, with the help of her
faithful followers she constructed a myth about the origins of Christian
Science that denied her dependencies and accentuated the uniqueness of
her discovery. But as we will see, those attracted to Christian Science
were primarily drawn not by Eddy or her story, but by the experiences
generated by the truth of Science.

2

Becoming a Practitioner and Teacher

Four years ago I learned for the first time that there was a way to be healed through Christ. I had always been sick. . . . Still, I thought if the Bible was true, God could heal me. So when my attention was called to Christian Science, I at once bought SCIENCE AND HEALTH, *and began to improve my health . . . until I found I was healed, both physically and mentally. Then came a desire to tell every one of this wonderful Truth. I expected all to feel just as pleased as I did; but to my sorrow none would believe. Some, 'tis true, took treatment and were helped, but went on in the old way, without a word of thanks. But still I could not give up. I seemed to know that this was the way, and, I had rather live it alone than to follow the crowd the other way.*
—MRS. FLORENCE WILLIAMS, Le Roy, Michigan, 1891

In an exceptionally complete *Christian Science Journal* testimonial, Mrs. E. D. S. recounted the story of her March 1888 conversion to Christian Science and the beginning of her healing career. Disagreeable experiences with regular medicine early in life had led her to become "disgusted with drugs," and she turned instead to the principles of hygiene, which she practiced for twenty years before coming to doubt their ability to prevent disease. In the terms of her later Christian Science beliefs, because of these doubts she "rapidly grew worse in health" until her sister gave her a copy of *Science and Health*. After she "had read to, and through, [the chapter titled] 'Teaching and Healing,'" she became so interested that she "began reading that blessed chapter over again" only to find that "I was cured of my dyspepsia, and that I could use my strength in lifting, without feeling the old distressing pain in my side. . . . Then I began my first conscious treatments." Thrilled again by her memories, she continued her narrative with words that tumbled forth:

A cry for help, knowing it would be answered; precious texts from the Bible, which had already become like a new book to me; sweet assurance of faith by the witnessing Spirit; strong logical conclusions, learned from SCIENCE AND HEALTH; what a wealth of material! Before finishing the book, all tendency to my old aches and pains had left me, and I have been a strong, healthy woman ever since.

After her own healing and even before she had completely read *Science and Health*, she cured her husband after just a ten-minute treatment by "demonstrating over" "the belief of bilious fever." Later that same year she "met a lady in our Rocky Mountain berry patches who complained so bitterly that I felt compelled to offer her treatment"; her first effort "cured" the woman. Excited by this rapid success, the author anticipated the time when she could "teach others the way of Truth, as well as add to the many demonstrations God has given me."[1]

Although the *Journal* rarely featured such a comprehensive account as this of a Scientist's spiritual biography, details like those in this story often reappeared in the hundreds of testimonials published in a more sketchy form. Representative of many who joined the ranks of Christian Science, Mrs. E. D. S. followed a path from personal healing to healer of family, friends, and strangers and finally to teacher of others.

But Mrs. E. D. S. also typified another characteristic of most Christian Scientists: she was a woman. As an 1887 observer put it, Christian Science's "apostles are mainly women. It is spreading rapidly and widely, and seems determined to stay. Its converts are mostly women. Its enemies are mostly men."[2] By 1906 women composed 72.4 percent of the Christian Science membership, whereas Congregationalists and Seventh-day Adventists, the American denominations with the next largest proportion of female members, contained only 65.9 percent and 65.2 percent, respectively, and the average for all denominations was 56.9 percent. Even more remarkably, the ratio among full-time Christian Science practitioners was five to one female by the 1890s and eight to one by the early 1970s.[3] What can we reconstruct of that always individual but nonetheless communal pilgrimage of Christian Scientists from nonbelievers to practitioners and teachers? And why did women so disproportionately outnumber men in the movement and dominate the ranks of practitioners?[4]

"They Wanted to Get Well"

In principle, Christian Science addressed all human needs, but interested persons usually first tried it because, in the words of historian Penny Hansen, "they were sick and they wanted to get well."[5] Most who became Christian Scientists did so when they, a friend, or a relative experienced a physical healing through Christian Science. But Christian Science also offered practical solutions to social, economic, intellectual, and psychological disorders and even claimed to heal sick pets, improve bad weather, and abolish bad habits.[6] Over and over, Christian Scientists acknowledged this primary role that physical healing played in winning converts. In an 1894 speech to the Woman's Council in Minneapolis, Minnesota, Mrs. M. A. Gaylord conceded that "as a rule people turn to the Science first, (and often as a last resort) to be healed of physical disease not knowing of its work along other lines."[7] Ezra W. Reid echoed her opinion in 1899: "Possibly the majority of those in Christian Science come through physical suffering and the healing thereof; another class (and this is increasing year by year), because of their dissatisfaction with prevalent religious teachings; still another, because of their dissatisfaction with Churchianity."[8]

In 1981 Penny Hansen, recognizing that two-thirds of the testimonies appearing in the *Christian Science Journal* from 1885 to 1910 related to female healing, analyzed them to explore this female connection. What did the testimonies reveal? After becoming disenchanted with their physical or spiritual health, prospective Christian Science patients consulted doctors first, not uncommonly over a dozen different ones. Their ailments covered a wide spectrum of physical and emotional ills: each sex possessed its own unique problems, but women often complained about childbirth. Typically, the patients had heard about Christian Science healing from another woman before they visited a practitioner, and after the mid-1890s they preferred practitioners of their own sex. Of all the persons healed by Christian Science (the testimonies reported only happy endings), a higher percentage of women than men joined the movement. Converts typically found a teacher, took instruction, and then began treating their own families and friends. Both before and after their first encounters with Christian Science, these women remained in charge of their families' health.[9] Of course Christian Science did not attract only women. Men also joined and came to wield power and influence within the movement far out of proportion to their numbers, but they always remained a minority.

Although Christian Science appears to have been theoretically well suited for success as a system of self-help domestic healing or as a religion of private contemplation, these testimonies revealed few such tendencies, despite the steady praise for the power of *Science and Health* to effect complete cures after a single reading. Even a superficial overview of the nineteenth-century health reform movement—with its supposed ties to women, their special sphere, and social reform and an acknowledgment of the health reformers' belief that doctors are not necessary because sickness can be prevented if nature's laws are only correctly understood and obeyed—exposes numerous parallels to Christian Science.[10] Nonetheless, for those who could afford them Christian Science healers proved to be not a luxury but necessary assistants for all who seriously sought continuous growth in truth.

A partial explanation may lie in recognizing that the testimonies that appeared in the *Journal* were not necessarily representative of Christian Science practice. Like most religious and medical testimonials, those in the *Journal* were selected by the editors to show Christian Science in the most positive light. In a surprisingly frank admission, one of the 1886 editors confessed that "we never do publish the tame and the ordinary. It would be dull, and nobody would read it."[11] At least twice Eddy advised editors of the *Journal* to curtail the number and types of testimonies. In 1898 she complained that there were too many testimonials; they appealed to a baser human nature, like narcotics and fiction.[12] And in 1905 she echoed her sentiments, concluding that too many pages of testimonials were like advertisements for patent medicines.[13] Furthermore, as Christian Science practitioners formalized their practices during the 1890s, they standardized prices, regulated education and training, and established codes of conduct. If anyone could heal herself by reading and applying *Science and Health*, then practitioners might have fewer patients, and their practices would suffer. The editors of the *Journal*, ever sensitive to the needs of practitioners, undoubtedly wanted to stimulate a market for Christian Science practitioners by making it appear that most healings occurred with their assistance. At the same time, they recognized that the best way to increase subscriptions was to keep practitioners happy, because patients subscribed when their practitioners "prescribed" the reading of "truth-filled literature."

The importance of practitioners, most of whom were female, to the healing process may also reveal something about the nature of Christian Science healing as a distinctly communal and feminine process. By making Scientists little less than gods, Christian Science theology may have appeared to reinforce in its members the masculine cultural ideal of independence and autonomy.[14] But by also encouraging them to seek out another person to assist the healing process, Christian Science guaranteed an interpersonal connection or healing relationship that implicitly acknowledged the communal nature of God and the cooperative nature of the search to rediscover one's identity with him-her. Moreover, the predominance of female "assistants," most of whom embraced Eddy's model of equal but separate spheres and rejected a defiantly feminist attitude toward gender relationships, promised women a sensitive and often empathic hearing, reminded men of their dependence on women, and reinforced for both a recognition that God is compassionate, not coercive, and cooperative, not domineering. Christian Science practitioners attracted women not just because they healed people or because they were women, but because as *female Christian Science healers* they spoke to women's distinctive ills in an affirming yet persuasive voice. When practitioners promised victory over sin, sickness, and death, those women who listened, listened intently because within the "woman's sphere," these three were the peculiar enemies they struggled against.

Cultural concerns for female delicacy created their own pressures on a woman's choice of practitioner, a fact many woman physicians exploited. When women visited female healers, they may have more openly discussed female diseases and complaints and matters related to childbirth. It would be wrong simply to accept apologists' tendency to reduce the attraction of the healing phenomena to simple metaphysics or to discuss healing only at its supposedly deepest or most profound level. Who is to say that the practical, physical consequences of healing were less profound or meaningful than the supposed deeper significance of ontological harmony, especially when much of the evidence we have suggests that most Christian Scientists were first attracted by the report of physical healings and the promise of like cures?

Appeal to "Troubled Souls"

Although the metaphysics of Christian Science healing may have received more than its share of attention, it does furnish an important perspective on the self-identity of the movement's members. Christian Scientists have long maintained that healing entailed more than simple physical recovery; in its most profound manifestations it involved a complete spiritual and physical conversion. As Christian Scientist and historian Stephen Gottschalk has asserted, "Christian Science made its deepest appeal not so much to sick bodies as to troubled souls."[15] Many modern scholars outside the tradition who have studied the issue have agreed, one result being that they more often have understood Christian Science healing as a variety of religious or spiritual healing than as a branch of mind or mental healing. Although the complex nature of healing makes such distinctions problematical, the evidence remains clear that Christian Science attracted many who searched for persuasive theology and spiritual retreat, not just for cure of disease. In fact its distinctive, metaphysical answers to serious religious questions may have proved especially attractive to women.[16]

Here again the Christian Science testimonies that appeared in the *Journal*, despite their doubtful representativeness, prove useful.[17] After a very cursory yet suggestive sampling of those testimonies, one scholar correctly concluded: "If we accept these converts' own words and the genuineness of their existential and religious dissatisfaction with the nineteenth-century theological status quo, then we have to concede to them a genuine religious motivation at least equal to the other motivations put forward to explain their attraction to Christian Science."[18] Contemporary observers agreed, but they often spoke in less charitable terms. They referred to women's emotional natures, which made them vulnerable to "various fads and 'isms,' and the quickest to credit all things to the supernatural," and one physician even suggested that Christian Science attracted women who had a "mental aberration" that led them to "live the combination of ordinary mundane life and an 'ethereal' one which constantly obtrudes itself."[19] However, contrary to such beliefs, often held by male critics, women converts to Christian Science were not just emotionally dissatisfied or mentally unstable; often, genuine intellectual dissatisfaction moved them. Mary Collson, C.S.B., a practitioner from 1904 until her resignation from the Mother Church in 1932, was attracted not only by the peculiar intellectual air of Christian Science metaphysics, which reminded her of giants like Plato and Ralph Waldo Emerson, but also by its claims to mathematically exact and demonstrable propositions—in short, its claim to be scientific.[20] Unpersuaded by traditional Christian theodicies, many asked, "How can a good God allow suffering?" Disappointed by the apparent failure of conventional beliefs to consistently alter behavior, they wondered, "When will Christians demonstrate the truths of Jesus in their lives?" And many, like Collson, believed that science in some form held the promise of true and practical answers.

For many nineteenth-century American women it may have been, as one historian has claimed, that the "disasters of childbirth, illness, death, [and] loss of security through recurrent financial crises" turned "thy will be done" into a "special female prayer."[21] But many other women, stimulated by the demands of their supposed special duty to maintain America's spiritual and physical health, found answers to these human dilemmas both emotionally and rationally inadequate and sought alternatives. Some threw off the real or imagined bonds of "true womanhood" and pursued a new ideal of "evangelical womanhood." Fashioned, according to historian Anne M. Boylan, by antebellum women interweaving "the traditional Protestant ideal of the 'vertuous woman' with a new evangelical stress on action," "evangelical womanhood" defined a sphere for women's activity in and for society as important as that of men. These women worked for the advance of Christian civilization by rearing good children, devoting themselves to their husbands, and establishing religious and social institutions, including benevolent associations, foreign and domestic missions, and Sunday schools.[22] As we have seen, such ideals fed the ambitions of the young Mary Baker Eddy and stimulated her efforts to reach beyond the constraints of her rural New England world.

As the nineteenth century moved past its midpoint, some American women gave up on religion altogether; but more typically late Victorian women continued their efforts to revitalize old denominations or, when such efforts proved fruitless, to form and strengthen new organizations. Such was the case with Eddy and her followers, who discovered that Christian Science offered a powerful response to perplexing, perennial questions. In fact, the three often repeated evils of sin, sickness, and death for which Christian Science offered solutions were precisely the three great evils against which Victorian women were to do battle from their "proper sphere." Carol Norton, one of the first Christian Science public lecturers, bragged that

> the work of Mrs. Eddy has opened to woman in the ministry of Christian Science, the two noblest of all avocations, philanthropy and medicine. Through the understanding of Christian Science men and women, by one and the same method, can reform the sinner and heal the sick. . . . By years of patient toil she [Eddy] has formed a system of religious and medical instructions that has already become a boon to thousands of mothers, because of its demonstrable power to strengthen moral character, and instill a natural love of the pure and good in the minds of children, and because of the freedom that it brings to families, inasmuch as it heals all manner of disease, destroys the fear of parents, and thus becomes the ever-present friend, the Guardian Angel, in the home.[23]

When women met defeat in their culturally imposed battles, many religious denominations advanced only comforting words of encouragement and theological rationalizations that often wore thin. But the Christian Science theodicy radically denied the reality of sin, sickness, and death, thus promising victory over all the vagaries and disappointments of life, and dispensed metaphysical solace for many frustrated and restricted Progressive Era American women.[24] In a not atypical reaction, Collson recalled, "[I] submerged myself" in the teachings of Christian Science "in order to escape from my world, which had become too difficult a proposition for me."[25]

The Christian Science doctrine of a God who exhibits both male and female characteristics, whose nature humankind ideally emulates, also proved attractive to many women, who often found that in their associations with men their "feminine" characteristics were stifled and their relationship to husbands was subservient.[26] During the last years of her leadership Shaker founder Ann Lee (1736–84) had planted the seeds for what became the distinctive Shaker doctrine of a Father-Mother God, but it was Mother Ann's nineteenth-century followers who used the doctrine to defend a social model of "separate but equal" treatment for men and women.[27] Many spiritualists also communed with a Mother-Father God and used their theology to buttress participation in the women's rights movement.[28] Despite Eddy's resemblance to mediums, however, she rejected their cry for gender equality and instead used her similar doctrine of God to justify her own view of gender roles. Developing the principle of "separate spheres," Christian Scientists repudiated a hierarchy between the spheres and redressed the often unequal relationship between husband and wife by drawing on the balanced characteristics of the deity as the ideal model for the marriage relationship.[29] According to Eddy,

> a union of the masculine and feminine mind seems requisite for completeness; the former reaches a higher tone from communion with the latter; and the latter gains courage and strength from the former; therefore, these different individualities meet and demand each other, and their true harmony is oneness of Soul. Woman should be loving, pure, and strong. Man, tender, intellectual, controlling. . . . [However,] man should not be required to participate in all the annoyances and cares of domestic economy, or woman to understand political economy; but fulfilling the different demands of separate spheres, their sympathies may blend to comfort, cheer and sustain each other, thus hallowing the copartnership of interests and affection whereon the heart leans and is at peace.[30]

This view of a generic human, ideally composed of both feminine and masculine traits, anticipated in many ways the opinions in Elizabeth Cady Stanton's *Woman's Bible* (1895, 1898) and proved popular among many women.[31]

Whether because of troubling theological questions answered by "Scientific" certainty or the resolution of long-standing gender inequities, many converts found a spiritual home among Christian Scientists. And when demonstrations failed and success proved elusive, Christian Science metaphysics, hovering in the wings, swept in to provide comforting reassurance. As anthropologist Margery Q. Fox has observed, "A world that offers little in the way of profit or personal fulfillment can always be repudiated as unreal. True progress and personal growth are then measured in spiritual terms."[32]

"Practical" Advantages

Despite the importance of understanding the attraction Christian Science healing held for those searching for physical or spiritual well-being, one must not underestimate the social and economic advantages gained by many Scientists who acquired an education and established successful healing practices. Mary Baker Eddy's boundless energy and her ambition for spiritual, literary, and professional success had proved as much, and her example effectively drew many women to careers in mental healing. As one contemporary critic put it, "Many women take up Eddyism as a business in the same way that they would take up midwifery, nursing, or massage—only Eddyism is more 'genteel' and more remunerative. Any one may obtain a Christian Science diploma after taking a course of twelve lessons for $100.00! For self-supporting women it is a godsend."[33] Although healings and metaphysics had attracted Mary Collson to Christian Science, she soon found that its "metaphysical appeal . . . ceased to excite her as much as did the possibilities for its practical application." When she "put out her shingle" in Amesbury, Massachusetts, during the winter of 1904–5, "clients, most of them middle- and upper-class women, began to stream in."[34]

With the exception of Eddy's earliest students, who came primarily from the working classes of Lynn, Massachusetts, Collson's experience was not unusual.[35] If one uses occupation as a measure of class, by 1910 most Christian Scientists were probably members of the "new" middle class, with about 51 percent professionals, 31 percent proprietors, managers, officials, clerks, and similar workers, and only 16 percent skilled, semiskilled, and unskilled workers.[36] Further confirmation of the upwardly mobile and generally middle-class status of most Christian Scientists comes by identifying what they thought they were (whether they had actually achieved it or not) and noting how they behaved. Christian Scientists emulated the postbellum middle-class attachment to institutions and titles as a means of embellishing their status. They not only founded the Massachusetts Metaphysical College and numerous Christian Science Institutes but readily assumed self-important titles such as "Doctor" and attached to their names educational degrees such as C.S.B. and C.S.D. Scientists built their careers as practitioners in part on what historian Burton Bledstein has characterized as "the prestige of white-collar employment that took pride in its competent service and its basis in science" and on their expertise in handling an esoteric body of Christian Science knowledge. Like Bledstein's "consummate Mid-Victorian individual," a Christian Scientist "never lost faith in the promise of his 'becoming,' despite adversity. He never gave up on 'making it.'"[37] In Christian Science practice, a practitioner simultaneously fulfilled her aspirations for service and acquired the sense of independence and pride that result from meaningful work.[38]

To further understand why Mary Collson's work as a practitioner may have excited her, we need to understand that for generations cultural and social restrictions had prevented American women from receiving public or professional credit for their crucial contributions to the "doctoring" of families.[39] The nurturing, mothering role defined for a woman by the "cult of true womanhood" virtually destined her to care for both the emotional and the physical ills of her family, but often at the price of denigrating her intellect and devaluing her contribution to the health of Americans. As scholars have observed, while "medicine in the nineteenth century was being drawn into the marketplace," healing became "detached from personal relationships to become a commodity and a source of wealth in itself," in short, "a male enterprise."[40]

During the mid-1800s, however, this state of affairs began to change as women mounted a defense of education and professional advancement for themselves either grounded in the values inherent in the cult of true womanhood or based on competing cultural feminine ideals.[41] Although doubtless pervasive, the cult of true womanhood was, as others have persuasively argued, only one of the "popular ideal[s] for middle-class American women [that] existed and was embraced between 1840 and 1880." "Real womanhood," with its ideals of competence, "physical well-being, vigorous health, and physical fitness" competed as one alternative to the fragility, pious self-sacrifice, and fashionable illness of "true womanhood." In fact, authors of prescriptive literature that upheld "real womanhood" urged "women to get formal medical training as physicians much more often than they urge[d] them to become nurses."[42]

Building as well on the record of their strong public leadership in the social causes of abolition, peace, temperance, and morals, women molded the public's view of "true womanhood," built upon the cultures of "real womanhood" and "evangelical womanhood" and championed the cause of health.[43] More and more women assaulted social conventions, gained access to medical education (often sectarian), and experienced the rewards of professional independence. Nonetheless, despite such advances, which led historian Gloria Moldow to observe that "in no profession except teaching did women have better prospects than in medicine," numerous cultural and economic restraints operated to curtail women's access to the practice of regular medicine.[44]

Among the cultural factors hampering some women's embrace of regular medicine lay their distaste for the theories and therapies of the regulars. In response they sometimes turned instead to the often friendly sectarians.[45] Some early nineteenth-century women expanded their domestic use of botanics into full-fledged Thomsonian practices and joined their male cohorts who challenged regular medicine and forced the repeal of most medical licensing laws.[46] During the 1850s the eclectics, liberal cousins of Samuel Thomson and his students of botanical medicine, for a time actively and successfully recruited women to one of the five largest medical colleges in America, the Eclectic Medical Institute of Cincinnati, where among the more traditional courses a student received instruction in the diseases of women and children.[47] The mid-nineteenth century witnessed a small vogue in hydropathy, which not only offered a natural alternative to regular therapeutics but also presented women with both a system of domestic care and professional advancement within the socially acceptable parameters of woman's sphere.[48] Women contributed significantly to the growth of homeopathy among middle- and upper-class Americans, and there is some evidence that women patients flocked to receive its gentle therapies.[49] Joining these medical sectarians in the late nineteenth century, metaphysical healers rose in popularity by attracting many female patients and offering them training as metaphysicians. By the mid-1880s Eddy's Christian Science stood out as one of the most widely known schools of the new mind cure craze and as more than just a marginally viable alternative for women seeking a healing career. Of the three women in the first medical class at Johns Hopkins, one graduated with honors, one dropped out to marry, and "one left to become a Christian Scientist."[50]

Not only did the evolving cultural values and economic structures of nineteenth-century America contribute to a growth in the number of female physicians at the turn of the century, they also inclined women to become Christian Science practitioners. Women were influenced by the women's rights movement, the persistent cultural connection between women's "nurturant qualities" and their traditional role as healers, standards of Victorian propriety that led them to shy away from male midwives and obstetricians, the medical sects that sometimes welcomed women in their battle with regular physicians, and the search for job opportunities.[51] Nonetheless, for most Christian Science practitioners their distaste for the drugging of regular medicine and their successful experiences with Christian Science, both spiritual and physical, seemed to have been the decisive factors in their decisions to join that particular school of healing.

Practitioner Training

All Christian Scientists practiced healing by "demonstrating over false claims" (that is, curing sickness and sin), but those who sensed a special calling and heeded Eddy's encouragement to devote themselves professionally to full-time service opened healing practices. As Scientist William R. Rathvon declared in 1900, "opening an office and hanging out a sign does not make one a Christian Science practitioner in the true sense of the term. A majority of those whose whole time, energy, and abilities are today given to the practice of Christian Science in healing the sick and sinful, did not seek the work. It sought them."[52] Once called, practitioners proved their worthiness by exhibiting an unshakable confidence in the teachings of Christian Science and an ability to remain convinced of the unreality of sin and disease even in the face of a patient's often vivid report of moral or physical problems.[53] They could not, of course, give evidence of any obvious defect in themselves that they had not "demonstrated over" (cured), but neither did they have to claim perfection.

Many practitioners learned their Christian Science by studying the Bible, *Science and Health*, and Eddy's other writings on their own and practicing on their patients. But for $300 students could take a Primary course of twelve lessons taught by Eddy or one of her students and thereby earn the right to call themselves Bachelor of Christian Science (C.S.B.).[54] Eddy encouraged class study. When an inquirer wrote to the *Journal* to ask, "If one is obliged to study under you, of what benefit is your book [*Science and Health*]?" Eddy replied: "Why do we read the Bible, and then go to church to hear it taught?—only that both are important. Why do we read Moral Science, and then study it at College? You are benefited by reading 'Science and Health,' but it is more to your advantage to be taught it by the author."[55]

Given the extraordinary effort by some to travel long distances at no little hardship to attend an Eddy course, apparently Eddy's invitation did not go unheeded. A former student's 1886 letter of praise for Eddy's abilities as a teacher hinted at what they experienced. "From hearing Mrs. Eddy preach," the author began, "from reading her book (however carefully), from talking with her, you do not get an adequate idea of her mental powers unless you hear her also in her classes. Not only is she glowingly earnest in presenting her convictions, but her language and illustrations are remarkably well chosen. She is quick in repartee, and keenly turns a jest upon her questioner, but not offensively or unkindly. She reads faces rapidly."[56]

Providing advice and professional experience as a sort of midrash on the text, the instructor led students through a careful study of key sections of Eddy's published writings. Ezra M. Buswell, a student in Eddy's May 1887 class, recalled that the "course was not for the purpose of teaching you how to heal the sick" but "for the purpose of teaching me how the sick were healed."[57]

Just before departing Lynn, Massachusetts, for Boston, Eddy obtained a charter for the Massachusetts Metaphysical College, which opened in Boston in May 1882 as an educational base for her movement. Although she touted it as "the first legally founded and thoroughly metaphysical institution in the world," the legal "agreement of association" signed by Eddy and six associates in 1880 more ambiguously claimed that the institution would "teach Pathology, Ontology, Therapeutics[,] Moral Science, Metaphysics, and their adaptation to the treatment of disease."[58] Eddy assumed a new academic title, "Professor of Obstetrics, Metaphysics, and Christian Science," and published a list of cooperating physicians in the Boston area, but in fact she was the institution's only instructor for its early years.[59]

According to the college's 1884 bulletin, the school embraced "the following essential characteristics of a radical reform in sanitary education":

First. The requirement that the candidate for admission present not only the intellectual and moral qualifications, but evince readiness to become a humble and intelligent advocate of the teachings of Christ Jesus as contained in the Scriptures and elucidated at this Institution.

Second. The provision for a thoroughly directed course of instruction of three weeks, and covering a minimum of three scholastic years of practical and evangelical healing before graduation.

Third. The demand that all students who receive degrees shall be examined at least once annually during the period of three years, either publicly or privately, as may be determined by the Board of Instruction, and furnish an original thesis on metaphysical practice.

Fourth. The total disregard of sex disabilities [replaced by "distinctions" in 1886] in student, teacher, or preacher.[60]

Although course content remained a world apart, such a course of study bore some similarity to the midcentury medical school requirements of a brief course of lectures and a three-year apprenticeship. By 1886 the curriculum included courses titled "The Principle and Practice of Christian Science, or Mind Healing," "Mental and Physical Obstetrics," and "Scientific Theology"—all helpful in providing the training in doctrine and healing technique necessary to establish a Christian Science healing practice. A student needed only two textbooks: the Bible and *Science and Health.*[61] Though graduates now received diplomas from a state-chartered school, their education closely paralleled the courses Eddy previously had taught privately since 1875.

In order to settle a student into the truth and confirm her orthodoxy, the college, for $100 tuition, offered its first Normal course in August 1884. Graduates of this course, usually offered only to mature students with three years of healing experience, received the Doctor of Christian Science (C.S.D.) degree, although many students called themselves Doctor whether they took the Normal course or not.[62] Many Normal course graduates followed Eddy's advice and established regional institutes across the country to spread Christian Science. Rapid growth of Christian Science activity in the states of Iowa and Illinois, for example, followed the founding there of sixteen such institutes during the 1880s and 1890s, before Eddy centralized educational control in Boston and ordered such schools closed at the turn of the century.

Beyond these two levels of instruction, between 1887 and the college's closing in 1889 some students enrolled in a special $200 course in metaphysical obstetrics, which applied the principles of Christian Science to the pain and associated errors of childbirth (see chapter 4).[63] In explaining why he took the obstetrics course, Nebraska practitioner Ezra M. Buswell stated, "My idea was that we might handle that question [birth] as well as any other in allaying the fear of the mother and those attending."[64] Finally, for those students who had received the lower degrees, practiced correctly for three years, "gained a thorough understanding of the Spiritual signification of the Scriptures" by completing the course in Scientific Theology and consistently lived "in accord with Christian Science," the College conferred its highest degree, the Doctor of Divine Science (D.S.D.).[65] Some one thousand students received training at the college during the seven years it remained in operation.[66] Armed with one or more of the college's degrees, Christian Science practitioners and teachers fanned out across America to bring knowledge of the Truth that heals sin, sickness, and death.

Establishing and Operating a Practice

Fresh from class study with Mrs. Eddy, Josephine Tyter, C.S.B., arrived in Richmond, Virginia, in July 1887 to bring Truth to a community in which she knew no one "and no one knew me."[67] Setting up a practice where "they knew nothing of" Christian Science had its advantages—initially the local physicians hardly noticed her. But that soon changed. At a meeting of the Tuesday Evening Literary Club a physician publicly denounced Christian Science with such vigor that one of Tyter's male patients felt constrained to rise in her defense. Believing that "it is not words but works" that one should seek "as a reward for your teachings," she avoided such public debates. Instead she devoted her efforts to healing and to preserving the purity of Christian Science against the encroachments of the ubiquitous mind curers. As a result of her labors she witnessed many healings, including the recovery of a pregnant woman who had "suffered pain constantly" and "had lain three months in bed," and she successfully "held the people on the right side" against mind cure. Despite such triumphs, however, she reported that "this city is the hardest place I ever tried to establish Christian Science in; but now it is done, and the harvest is great, as the majority of the people [in mental healing] are for it [Christian Science]." Looking back on her first year of isolation in Richmond, she testified that "I have seen but one Christian Scientist since I was in Boston, but I do not feel lonely or weary, for I had such a longing desire to conquer this city with Truth, that I would not mind if I were in a desert."[68]

In many ways Tyter's experience typified the experiences of practitioners who entered new territories with their teachings and practices.[69] Wherever they went, they targeted urban areas. As Christian Scientists moved westward, they continued this pattern, so that by the turn of the century the rapid midwestern growth of Christian Science coincided with the large numbers of women migrating to major cities, such as Chicago, in search of work.[70]

After Christian Science healers arrived, they posted signs outside their homes advertising themselves as Christian Science practitioners, took offices in the business districts of towns, boldly printing their names and titles on the doors, and distributed or inserted in the local news-papers business cards listing their business addresses and office hours with the title "C.S.B.," "C.S.D.," "Christian Science Teacher and Healer," or "Christian Science Dispenser" after their names.[71] After their presence in the community became known, such promotion became unnecessary as news of their activities spread from patient to patient by word of mouth. Soon after George W. Wheeler arrived in St. Joseph, Missouri, the fol-lowing advertisement appeared in the local paper:

It is a fact worthy of notice that christian science healing is gaining in popular favor very rapidly, and all because of its success in accomplish-ing what it claims. Among the many scientists in St. Joseph we take pleasure in noticing the increase, as well as the success of the practice of Dr. Geo. W. Wheeler, who has been here only a few months and who is located in very pleasant office rooms at No. 820 Francis street. The princi-ples of christian science recommend themselves to the careful thought of every one, and Dr. Wheeler has been very thorough in the study of christian science mind healing and is always ready to give information upon this subject that is now engrossing the attention of all thought-ful people.[72]

When Samuel Putnam Bancroft attempted to establish a practice in Cambridge, Massachusetts, during the winter of 1874–75, he first had to determine a title for himself and a location for his office. "The location was left to me, and I was told later that I had made a poor selection; that I should have located nearer the colleges; would have gained prestige, etc. The name, under which I was to practice was determined by Mrs. Eddy, and my [business] cards and sign read 'S. P. Bancroft, Scientific Physi-cian, Gives no Medicine.' Advance notice was undertaken by a distribu-tion of cards, circulars, and a few letters."[73]

During the 1880s many practitioners carried out these efforts to establish and advertise their practices with an all-consuming passion. According to one student's recollection, Christian Science teachers exhorted "us that if we did not devote our whole thought and time to Christian Science healing of the sick, we could neither keep the understanding gained, nor continue in health or demonstration for others." Some leaders even urged mothers to abandon their children so they could more rapidly spread the message. Apparently such zeal came under criticism, however, for the student concluded that "this, many of us *have demonstrated* to be a mistake on the part of the Teachers themselves."[74]

Some practitioners, and not just graduates of the Normal course, took a less aggressive approach to advertising their practices and felt no tension between their call to mission and their choice of profession. Viewing themselves more as religious teachers than as professional practitioners, they witnessed for a restored Christian gospel and expected Americans to accept them as they did the professional clergy. Emma S. Davis introduced southern California to Christian Science in 1887 with the first public healing in Riverside. She recalled that the second day after she arrived in Riverside "one of the proprietors of a newspaper called with his note-book in hand, asking me if I would give him some particulars regarding my purpose in coming here; as I doubtless wished to bring my business before the public. I told him I would talk with him if he would lay aside his note-book, but not with any idea of his mentioning what I said in his paper, as I did not come here with business, nor looking for business."[75]

When another reporter pressured her further the next day, she forcefully responded that "the Christian Scientists have but one method of advertising. 'By their fruits ye shall know them.' "[76] Ezra M. Buswell, pioneer Nebraska practitioner, echoed Davis's sentiments when he stated to the court at his 1893 trial for practicing medicine without a state certificate that his "business or occupation" was "a Christian Scientist so far as I understand and live up to the teachings of Jesus Christ."[77]

Both Davis and Buswell viewed themselves as religious healers, but not ones "in the business" of healing. If the testimonies published in the *Journal* provide an accurate indication, before 1889 the movement generally tried to attract attention by highlighting the large number of healings a practitioner could expect. For example, in 1887 Emma A. Estes bragged that on a visit to Newark, New York, "I was kept busy everyday.

Had forty-nine patients [during my three-week stay], and found my work greatly blessed." And Emma D. Behan, founder of the work in Kansas City, reportedly "treated about five-hundred patients" during the first year of her work in that city.[78] After 1889, however, the movement shifted attention from the number of healings to the success of teachers and practitioners in directing classes and founding churches. Of course healing continued, but this hybrid nature of the Christian Science practitioner—both religious teacher and medical healer—confused a society growing to identify "treatment" for physical ills with physicians and "treatment" for spiritual ills with the clergy. Even more confusing, few if any clergy established fee schedules for pastoral visits as Christian Scientists had done for their "treatments" by the turn of the century.

A wide variety of receptions awaited practitioners on their arrival in town or country. Most often the physicians and clergy of a town raised the biggest ruckus, while the townspeople ignored the situation unless an epidemic threatened a neighborhood where Christian Scientists lived or a child died under Christian Science treatment. Little, of course, was more troubling to Scientists than the often heated reactions that accompanied the criminal investigations or indictments that often followed such incidents. But positive receptions occurred as well. Soon after receiving instruction from Eddy, Julia S. Bartlett left Lynn, Massachusetts, in 1881 to establish Christian Science in Boston. At first the work progressed slowly, but by 1882 her practice "had increased to about thirty patients a day which, with church work and whatever came up to be attended to, made a busy life." In 1884 Bartlett healed a young woman who came to her from New Hampshire. After the patient returned home, many in her area wrote to Bartlett and requested that she visit the region. Finally she agreed, stayed for eleven days, preached and healed, and treated seventy patients a day.[79] She wrote to Eddy:

> I do not believe any student ever had such an experience as I am passing through. There is a perfect *rush* of patients. Three M.D[.]s are sending me patients at the same time some of them are terribly frightened in respect to their own position. They say they never saw such a string of people as are going to see that Dr —— [Bartlett.] I am turning away ten or a dozen patients every day that I cannot find the time to see. . . . Shall sell about 80 books [*Science and Health*] here and have 70 subscribers [to

the *Christian Science Journal*].... I am just about earning money from my practice at the rate of over $200 per week but that seems of no consequence compar[a]tively.[80]

Promotion to Teacher

Itinerant Christian Science teachers who roamed the country planting their seeds of Truth and reaping a harvest of practitioners, and their more permanent fellows who cultivated healers through the Christian Science institutes of the 1880s and 1890s did the most to hasten the spread of Christian Science during the decades before and after the turn of the century. Although the westward migration of settlers played a minor role in the diffusion of Christian Science, geographer Ary Johannes Lamme III concluded that Christian Science expanded from 1875 to 1910 predominantly through the work of teachers and the publication of literature.[81] These teachers, increasingly Normal course graduates after 1884, primarily used the publicity and proselytizing methods of religious evangelists in their efforts to attract students and establish themselves in a community. They followed the example of itinerant antebellum Protestant missionaries before them, who preached, taught, distributed literature, and established churches and schools. These methods proved as effective in the postbellum new West as they had in the old West, in part because, as historian Ferenc Szasz has suggested, so many of the conditions remained unchanged: "the wide variety of peoples, the frenzied pursuit of riches, the creation of instant communities, and the general collapse of social restraints."[82] Christian Scientists advertised public lectures or church services in newspapers, through handbills and broadsides, or door-to-door and distributed or sold literature, especially *Science and Health* ($3 a copy), throughout their communities. Every convert considered herself a missionary committed to teaching and ready to distribute *Science and Health*, not unlike Latter-day Saints with the Book of Mormon.[83]

Most of these Christian Science teachers were women, not men, and beyond the peculiar attractions that Christian Science may have held for them, they felt a call to witness not unlike the impulse experienced by many other nineteenth-century Protestant women. They wanted to spread the gospel, advance Christian civilization, and indulge a spirit of adventure. In return they gained a sense of doing something outside the home that garnered vocational security and prestige within Christian Science. Just like the women of the foreign mission movement, they had, in the words of historian Patricia Hill, bridged "the domestic sphere and the male-dominated arena of public life."[84]

Primarily through the work of these agents, membership in the church grew rapidly. According to the United States Census, in 1890 there were 8,724 Christian Scientists, but probably 1,500 at the most were orthodox followers of Eddy.[85] By 1906 membership in the Church of Christ, Scientist, had risen to 40,011, a stunning net gain of 2,500 percent.[86] For roughly the same period (1890 to 1910) two other indigenous American religious sects, Seventh-day Adventism and Mormonism, experienced only 125 percent and 142 percent net gains, respectively.[87] Although Eddy forbade the reporting of membership statistics after 1908, the number of licensed practitioners, which rose from 5,394 worldwide in 1913 to 7,828 in 1925 and 10,775 in 1934, gives an indication of the movement's continued growth into the early twentieth century.[88]

Christian Science spread rather slowly throughout New England from its home base in Massachusetts, but wherever it went it grew most rapidly in the smaller and medium-sized urban areas (10,000 to 100,000 population). By 1890 Christian Science possessed the strongest concentration of believers in the north-central region (57.6 percent of members), an area dominated by Yankee-Yorker migrants who generally followed the Mohawk River–Erie Canal–Great Lakes migration route westward to the growing urban centers of the Midwest and beyond.[89] Outside this region, New York (1,268), California (814), Massachusetts (499), and Colorado (447) contributed another 34.7 percent to the total church membership in 1890.[90] Based on the 1906 United States Religious Census, the percentage of the total Church of Christ, Scientist, membership for the north-central region and these four states had slipped from 92.3 percent to 79.1 percent, but it remained a substantial majority, with the north-central region containing 47.8 percent of the membership and the aforementioned states adding another 31.3 percent.[91] The South, cut off from the Northeast by war, culture, and its conservative, traditional Protestantism, rarely welcomed Christian Scientists until the late twentieth century brought the "snowbirds" to Florida and the Gulf.[92]

Christian Science attracted converts primarily because it promised to heal their bodies and souls. However, its distinctive theology and the "practical" social and economic advantages that followed a career as

Christian Science practitioner or teacher played an important secondary role in drawing women to the movement. Once trained as practitioners, these converts had open to them not only a career that promised economic security and social mobility but also a calling that filled their lives with a sense of purpose and direction. Although the paths most American women chose to achieve these ends differed from that of the Christian Scientists, many American women shared the frustrations and aspirations that set their Scientist sisters searching for greater fulfillment.

3

"Occasions for Hope"

Patients and Practitioners

I went inside of the office and a man came out of the inside office,—there are two offices. I asked if he was Dr. Anthony. He said he was. Asked me to step inside. I went inside, to his inside office. I told him that I had indigestion. He asked me if I had ever been treated before. I said no. Asked if I understood his method. Told him no. He said it was mental. Told me I had better buy a book for three dollars, which he had, but said that it was not necessary for a cure. Said he would cure me in three visits. Told me to eat anything I wanted to. Sat still for about ten minutes, looking at the floor. Looked up and said he thought I would feel better. He would like to see me again Monday night at his residence, 58 Governor Street. Charged me two dollars, which I paid. Said pay the balance, three dollars, on the third visit.
—Ernest Hall, witness at trial of Rhode Island
practitioner David Anthony, 1897

John H. Thompson, C.S.B., and his wife, Lida, joined the Mother Church in 1898 after reading *Science and Health* and other Christian Science literature during the early 1890s, when they lived in the state of Washington. After a short period of employment in the Eddy household at Concord, New Hampshire, John and Lida moved to Boston in 1899, where they worked as practitioners and attended the Normal class taught by Edward A. Kimball, C.S.D., in June 1902. With the exception of a year and a half (1900–1901) spent helping the Christian Science church in Fabyans, New Hampshire, the Thompsons remained in Boston to tend their thriving practices. Late in 1909, in order to give Lida more room at home for her practice, John moved his offices downtown to 372 Boylston Street, where he worked until his death in 1928.[1] The records that survive from Thompson's practice allow us to eavesdrop on a typical turn-of-the-century Christian Science practitioner as he rescued his patients from ill health.

Like his fellow practitioners, Thompson treated patients not only in person but also absently, which meant that he used the principles of Christian Science to treat those who were not physically present. Nebraska practitioner Ezra M. Buswell described the theory of absent healing when he testified at his 1893 trial for the illegal practice of medicine:

Q. How do you do it when not present?
A. We understand that mind [God] is ever present and that we can think of our patient and those suffering just as well if in England as we can in our own country and we understand that God will hear as well in Beatrice [Nebraska] as in Liverpool.[2]

Since God did the healing and he is omnipresent, there was no need for the patient to be present. In fact, according to Buswell, practitioners became involved with such cases only because the patient's awareness of his need of Christ could be expanded "through the understanding that we have of what God is and man's relation to him[;] I am enabled in some degree to express that to those who are suffering." That they became healed when they had not seen the practitioner was for Buswell "pretty good evidence that it is God that heals and not me."[3]

Fortunately, the carbon copies of some of Thompson's outgoing correspondence for the years 1908 to 1915 have survived, and they indicate that he maintained an extensive clientele of absent patients.[4] Rather than personally consulting with these patients, Thompson depended on letters and phone calls to keep up with their progress, to communicate necessary advice or treatment, and to bill them for services rendered. Although these letters do not reveal what Thompson actually did during an absent treatment, many do contain the advice and specific instructions that he tailored to individual patients. Most of these patients lived in New England, but some lived as far away as Australia and England; however, distance made no difference to the effectiveness of absent healing.

The letters varied in length, with Thompson's first letter to a patient often filling one and a half single-spaced typewritten pages. The subsequent letters often began with an acknowledgment of the receipt of payment for the previous treatments or a query or occasionally a chiding when payment had not arrived. Then followed updated advice and encouragement based on the progress of the case. After 1911 the letters were almost always less than half a single-spaced page. Some internal evidence hints that he may have become busier for some reason, possibly because he had more patients or pressing outside interests. The general advice in each letter included most of the same elements for every patient, but the letters were not simple form letters. Thompson personalized the advice in each to fit the patient's particular circumstances.

If these letters are representative of Thompson's practice, most of his patients solicited advice about some medical problem, although some asked for marital or business counsel, and occasionally someone raised a metaphysical question about the reality or unreality of evil. Thompson's 1910 letters to a Mrs. Dove, who, among other maladies, had an injured knee, illustrate the kind of issue he typically addressed with patients and the advice he often gave. He began his first letter to her, dated 11 June, with the following words: "Mrs. Kate Clark has asked me to take your case and I began working for you yesterday and should have written you then. I presume you know something of the teachings of C.S. and I will expect you to read Science and Health[,] that is the only medicine prescribed, on page 468[;] read and commit to memory lines 9 to 15 and whenever troubled or in any distress repeat those lines many times."[5]

As was his custom, Thompson first recommended that his patient concentrate on a particular passage in *Science and Health*—often, as in this case, the scientific statement of being—as a way of refocusing her attention from her problem to the ideal. Found in "Recapitulation," the summary chapter of *Science and Health*, the scientific statement of being stated: "There is no life, truth, intelligence, nor substance in matter. All is infinite Mind and its infinite manifestation, for God is All-in-all. Spirit is immortal Truth; matter is mortal error. Spirit is the real and eternal; matter is the unreal and temporal. Spirit is God, and man is His image and likeness. Therefore man is not material; he is spiritual."[6]

Scientists believed that this condensed statement of the central truth of Christian Science, when repeated like a mantra during difficult times, began to free them from the habitual errors of belief. Thompson tailored subsequent "prescription passages" selected from the Bible, *Science and Health*, other writings of Eddy, or authorized literature for the specific complaint of the patient. As he wrote to another patient, "The reading is the medicine prescribed by the Scientist as the M.D. prescribes material remedies."[7]

After his initial guidance for Mrs. Dove, to make his point clearer Thompson turned, as was his habit, to one of several illustrations, many adapted from Eddy's writings. He continued in his first letter to Mrs. Dove: "You will learn God did not make you to suffer, He sees you just as He made you, and He made you like Himself just as principle made 2×2 according to itself 4 and not 5 and always keeps it so, to say $2 \times 2 = 5$ would be a lie or false belief about 2×2 and a denial of principle."[8]

In this illustration, which Thompson used repeatedly with variations, he argued that just as one must understand that two times two always equals four and never five in order to perform consistently accurate mathematical calculations, so one must understand the underlying nature of reality before one can fully live a harmonious life in the real world of God's creation.

Part of that understanding came with the dawning realization that humans are not and never have been other than harmonious and perfect reflections of God. He reminded Mrs. Dove that during a dream we may feel we are in grave danger. When we awaken, however, we not only lose our fear but recognize that we never were in danger. So it is for Christian Scientists: the human experiences of sin, sickness, or death are like dreams from which we awaken to discover not only that we are perfect but that we never were otherwise.[9] On other occasions Thompson used a story about a tramp and a prince that bore a passing resemblance to Mark Twain's *The Prince and the Pauper* (1882). In Thompson's version, a tramp had stolen the son of a king and raised him as his own son. If you met the son one day while he was out begging, to remedy his problem you would only need to convince him that he was and always had been a millionaire and thus "put him in possession of his own."[10] Through both illustrations Thompson taught that healing comes through a realization that one is and always has been perfect.

Subsequent letters from Mrs. Dove reflected her disappointment that her healing was progressing so slowly. Thompson replied several times to cheer her and defend Christian Science in the face of its apparent failure. He sought to encourage her by reminding her that her "belief in constipation" had improved and that her "belief of lam[e]ness" must have improved as well because she reported that she had been "doing more work than common." He told her one of his cases had taken two years to cure, but that things had finally worked out marvelously, and he asked her to be at least as patient with Christian Science as she must have been with regular medicine. Returning again to two times two equals four, he reminded her of the importance of principle and warned her against looking at the body for evidence of improvement. "God made you well and without one affliction and He keeps you so," Thompson wrote, "and that is how He sees you, and that is how you must hold your own thought about yourself, never mind your knee but learn the Truth and hold to that and that helps you, and helps me to help you and will make you well, and besides being well you will know the Truth that will keep you well." He concluded with a brief reminder that she owed him $5 a week for her treatments and enclosed a copy of a Christian Science periodical along with the advice to read the testimonies of slow healing and avoid fixating on quick healing.[11]

In the last surviving letter to Mrs. Dove, Thompson again told a story to address his patient's concern that her knee did not seem to be improving. He said that the healing process is often like a person who, with only a compass to guide him, takes a trip from the East to see the beautiful flowers in California. As he travels he sees many flowers that confirm that he is on the right path, but at times there are no flowers, only mud and snow, and he must trust his compass to tell him he is going the right way. Nonetheless, he refuses to become discouraged and finally arrives at his destination. So it is with "the patient who seems to respond slowly, sometimes and quite often it looks as though he were getting farther away in place of growing nearer to health but with confidence in the Christ-Truth and Science and Health with the Bible to consult" he must persevere and show trust in God and in Truth, "and this trust will never be unrewarded." Thompson testified that in many similar cases of slow healing, as patients "steadfastly clung to the reading and tried to live what they found was true in the book, many things such as impatience, selfishness, anger, hate, malic[e,] revenge, carelessness, greed, av[a]rice, ignorance, superstition and in-numerable other things have been found to have vanished when the physical healing was reached, this is the physical healing, it is those false beliefs that make us sick and their destruction makes us well."

Thompson promised her that he would continue to work for her and remain ever confident that she was improving. He told her to think as little as possible about her knee, but instead to keep her "thought on what is true according to God who says you are in His likeness, hence you are spiritual and perfect, so be of good cheer, of good comfort, for the Comforter is near."[12]

Many scholars of Christian Science have focused on its religious beliefs and practices in an attempt to root the movement in the context of church history.[13] In so doing, however, they often have neglected the movement's affinities with medical practice and ignored the ways Christian Science practitioners offered patients religious healing couched in the forms and language of medicine. As the brief description of Thompson's practice illustrates, Christian Science practitioners engaged in a healing business that offered a therapeutic alternative to many patients for whom regular medicine had proved unsatisfactory.

Rare as well are studies of the movement that distinguish between the prescriptive pronouncements made by Eddy and other church leaders about how Christian Scientists should live and deal with their patients and the way they actually did so. Too often hagiographers have ignored any inconsistencies, however trivial, between belief and practice, while detractors have gleefully but often simplistically emphasized contradictions. The best glimpses of the practices of Christian Science practitioners have come from cultural anthropologists who have examined the practices of contemporary practitioners, but Christian Science practice during its vital formative years has been largely ignored by historians.[14] Therefore we do not need another prescriptive religious history of Christian Science healing. Instead we need a historical understanding of Christian Scientists' healing ritual that strikes a balance between a simple retelling of the Christian Science worldview and a critically informed translation of that culture to nonbelievers. What did Christian Science healers do, why did they do it, and how did their beliefs and practices change over time?

Patient History, Examination, and Diagnosis

Patients consulted practitioners for a wide variety of problems, from sickness and death to all sorts of sins and worries, including drinking, smoking, aging, overweight, marital tensions, and business problems.[15] Those who made Christian Science a way of life desired not only to be healed of pain or infirmity but to be purified. They thus, in the words of historian Grant Wacker, affirmed the "timeless bond between physical infirmity and religious anxiety" and "between physical health and religious assurance."[16] Practitioners exulted in treating patients whom physicians had apparently given up on because of chronic or incurable problems or for whom clergy had failed to provide spiritual and psychological satisfaction. According to practitioner Lloyd B. Coate's opinion read before the Present Day Club in Dayton, Ohio, on 29 January 1901,

> few people, if any, come to Christian Science for help until they have given the physicians a trial. A survey of the healing work accomplished in Christian Science in this city and near vicinity within the past five or ten years shows that almost every disease known in this section of the country has been successfully handled, generally speaking, including a large obstetrical practice. . . . We again reiterate that the physical healing is the smallest part of the work to be accomplished in Christian Science. The physical healing and moral reformation are one and inseparable.[17]

Unless it was an absent treatment, practitioners, like physicians, usually "advised" or "argued" (treated) at their offices (often rooms in their homes) or in their patients' homes. A witness in the 1911 trial of New York City practitioner Willis V. Cole described a typical two-room office suite with an outer waiting room and an inner treatment room in a downtown office building. The reception room "had two tables and a clock in it and [Christian Science] literature there, and ⌈about ten⌉ people sitting around, some talking and some reading." The inner room contained only a telephone, a desk, and several chairs.[18]

When patients first visited a Christian Science practitioner, the healer did not carefully inquire into their complaints to discover specific symptoms. This would have only strengthened the patient's "claim," a term Scientists often used to avoid confirming in their patients' minds the false or mortal belief (error or misconception) that their diseases or other problems were real. Instead, the practitioner usually attempted to get only a general idea of the "supposed" spiritual or physical problem so as to "act more understandingly in destroying it."[19] Emma S. Davis, C.S.B., who pioneered the Christian Science work in southern California, recalled in 1901 that when she worked among Spanish-speaking people a translator advised her of each person's problems. She recalled:

> Many came for help. A lady in the house counted between forty and fifty in one week. One day, hearing an unusual noise on the porch, I

looked out and saw seventeen standing there, who had come in two wood wagons which were before the door. One of the number had been previously healed. Occasionally he would speak a word that I could understand. He brought in a part of these people at a time, *making me understand their ills,* and when they went out others came in and he interpreted for them.[20]

Before she began her treatment, Davis wanted a general understanding of her patients' ills, but not a blow-by-blow account.

Because Christian Scientists viewed health in what might be called holistic terms, the practitioner often solicited additional background information about a patient's social circumstances. For such reasons Thompson asked one of his patients, a woman named King, "Please give me some additional information about your case unless head-ache is the main trouble; give me some of the things you seem to have to contend with every day; and something more about yourself, is there a Mr. King? and is he interested in Christian Science? and have you Scientists in your vicinity?"[21] While this patient thought of her problem only as headaches, Thompson suspected that the headaches hid deeper household difficulties, such as conflicts with an unbelieving husband. A true healing would involve more than removing a symptom; it would require resolving interpersonal tensions through a complete spiritual understanding. Thompson felt that in order to lead her to that understanding, he needed more particulars about her life.

Although the practitioner did not systematically examine a patient's body, she was neither blind nor deaf and doubtless observed a patient's condition, despite her best intentions to see the world only through Christian Science spectacles. In addition, both the practitioner and her patients undoubtedly absorbed some of the medical vocabulary ordinary Americans used to describe their health. Therefore, a practitioner could not completely avoid learning something about her patients' symptoms, either by simply observing them or by listening to the uninvited (but not always ignored) descriptions of patients, family members, or friends.

Consequently, language often got in the way of Truth. The following exchange, which took place during a 1902 grand jury investigation of practitioner John Carroll Lathrop in New York, illustrates the often confusing way such medical and "scientific" observations became intertwined for Christian Scientists:

Q.—What condition did you find Bessie in?

A.—I found her under the claim of what is called tonsil[l]itis.

Q.—What is tonsil[l]itis according to your belief?

A.—Merely a mortal belief.

Q.—What is cancer?

A.—I would say that cancer was a belief of the human mind made manifest on a human body.[22]

Assuming that Lathrop followed normal procedures in his treatment of Bessie, he would have been informed that she had a severe sore throat, but he would not have peered down her throat or palpated her tonsils. Nonetheless, he knew enough about the symptoms and the ordinary diagnosis of them to call them a "claim" of tonsillitis. As Eddy told a small group of her followers, "You would not argue diphtheria if the case was consumption; neither do you argue thunder and lightning when it is sin (malice)."[23] However, some extremist practitioners refused even to acknowledge the "claim" because they feared they would thereby reinforce it. In 1906 Eddy, specifically referring to death but with implications for all errors of belief, cautioned against such "straining at gnats" and invited practitioners to "definitely name the error, uncover it, and teach truth scientifically."[24] In other words, common sense dictated that one often must use ordinary language and commonly held beliefs in order to talk about a case, but at the same time one must always point to the true reality beyond.

Whereas physicians wanted to diagnose a patient's disease in order to apply a specific treatment, for Christian Scientists identifying a disease or naming it was not only irrelevant but dangerous. If a practitioner named a disease, it only increased the patient's fear, helped to confirm an error of belief regarding its presence, and delayed the cure. The following exchange took place between the prosecuting attorney and Milwaukee practitioner Crecentia Arries at her 1900 trial for the unlawful practice of medicine:

Q. That is to say, through your material senses you can tell in some cases whether a person is ill or not, can't you?

A. Yes, sir.

Q. And consequently you do consult your physical senses in determining whether a patient that consults you is ill or not?

A. No, sir, not physical senses.

Q. How, then, could you tell whether Miss Irma Grossenbach had any symptoms of illness before that morning?

A. She didn't show any. She looked the picture of health.

Q. She looked the picture of health?

A. (Not answered by witness.)

Q. So you do note things that come within your physical sense as to the health or illness of a person?

A. We can't help but seeing that much surely.[25]

What we see here is Arries's struggle to affirm what she believed to be the ideal (the true reality that all is well) while at the same time remaining honest with the fact that she observed with her own eyes the symptoms of health or disease. But she also simultaneously recognized that if she did not deny the reality of the disease and its symptoms she would further confirm the "claim" in both herself and her patient. These efforts perhaps seemed to be a foolish game of mental gymnastics to most unbelievers, but to a Christian Scientist they represented the essence of the struggle against mortal mind (error).

Christian Scientists believed, in the words of Thompson, that *"man is always well,* it [can]not be otherwise for all of God's creations are perfect and He keeps them perfect."[26] When Emma S. Davis testified for Eliza Ward in Ward's 1892 San Bernardino trial for manslaughter, Davis asserted that "we never diagnose any case. . . . Christian Science never cures without removing the cause; we will treat any disease no matter what it is; we would pray for a man if he took a dose of strychnine just the same as we would treat a sore toe."[27] For Davis the cure followed not from a correct diagnosis, but from understanding the underlying cause of all sin, sickness, and discord. If she had collected empirical evidence from her patients and categorized it according to medical theory, she would only have perpetuated their problems. Rather, she needed to unravel the skein of error that had bound her patients to their false beliefs, thereby setting them free. The general description and history she received from her patients allowed her to grasp the first thread.

"Argument": Treatment

After finishing her assessment of the problem, the practitioner began the treatment, or "argument," comprising a combination of silent and audible prayer and advice. According to early twentieth-century practitioner John Carroll Lathrop, treatment "is a realizing prayer, an enlightened faith, and a spiritual understanding of God, which is reflected by the practitioner, and reaching the consciousness of the patient, eradicates the belief in disease, which we consider is purely in the human mind."[28] Willis V. Cole testified in 1911 that treatment and prayer are identical: "I prayed—our whole interview took about twenty-five minutes. I remained all this time in silent prayer. That word prayer is a synonym for treatment."[29] Such "prayer," however, was not a believer's request that a benevolent, personal God bestow the gift of health on a sick person, but rather an affirmation of the Truth that God, the "Principle of man, and not a person, produces all good."[30] As such, Christian Science "prayer" only dimly resembled the petition for divine healing typical of most Christian traditions. In the words of former Christian Science practitioner and teacher Arthur Corey, "The church publication committee would have us refer to treatment as *prayer*, an effective device for winning respect in a predominantly Christian community and one which would give judges and juries pause in decisions against Christian Scientists for medical neglect. Who can afford to pass judgment which might be construed by an unsophisticated populace as an attack upon the right to pray for the sick?"[31] One need not adopt Corey's cynical tone to acknowledge that more often than not Christian Scientists on trial for their practices conveniently downplayed their idiosyncratic definition of prayer.

Typically, the practitioner positioned her chair before the patient, bowed her head in her hands or closed her eyes and raised her hands to her face, remaining in perfect silence for fifteen to thirty minutes. After receiving treatment from Rhode Island practitioner David Anthony in 1897, Ernest Hall recalled that the healer had "sat still for about ten minutes, looking at the floor. Looked up and said he thought" that Hall "would feel better." On concluding the session Anthony gave his business card to Hall. It read: "David Anthony, C.S.D. Room 10,—333 Westminster St. Residence 58 Governor St., Providence, R.I. Hours 1 to 4 P.M."[32]

It is difficult to assess what patients thought a treatment really was. Their understanding undoubtedly depended in great part on the extent of their knowledge about the teachings of Christian Science. Clearly some thought of the treatments as a kind of "spiritual medicine." Former patient Jennie A. Spead testified that even though she may not have understood what she read in *Science and Health*, she had relied on the words "because he [Tomlinson] had asked me to read them, and I took them as medicine, same as I would take a drug from a doctor that I didn't know anything about."[33]

Described from the perspective of the practitioner, the treatment took on a more elevated tone. David Anthony summarized his approach in the following terms:

> My method is to mentally and audibly inform the person who comes for help [of] his relation to God,—that it is impossible for him to have disease because there is none. Of course when they come they say they have some trouble. To show them that they have no disease, I explain to them that it is impossible and that is my only method. This I usually do mentally. I do it mentally because they would not understand it quite as well or would reject it more if I told them audibly. I simply show them their relation to God.[34]

At her 1900 trial for the unlawful practice of medicine Emma Nichols expanded on the rationale for inaudible treatment: "If the patient is sick and Christian science tells that patient he is not sick, you can see what position that would raise, for, to his senses he is sick and he is suffering, and, therefore, it would not be wise on the first visit to the patient to say to that patient, who is, perhaps in bed, and has been for weeks, that they are not sick."[35] Edward Everett Norwood, a late nineteenth- and early twentieth-century practitioner in Tennessee and South Carolina, offered readers of the June 1903 *Christian Science Journal* a slightly different explanation for silent treatment: "We must remember that when a patient comes to us for treatment, it is not a sick man coming to be healed, but a type of error coming up for us to cast out, and when we have destroyed all sense of that error in our own consciousness, our patient will be healed."[36]

Christian Scientists viewed healing essentially as the consequence of a rational conviction that sickness, sin, and death do not exist; the healer, because of her better understanding of that truth, served as a midwife or assistant who helped patients recognize their health by arguing that their symptoms and diseases did not exist. The practitioner sought to allay patients' fears by denying their problems and affirming that evil does not exist. Drawing on a simile to explain the practitioner's role, Thompson believed that "a practitioner is something like a music teacher to mentally or audibly show the way but the patient must act out this or practise it, and the error cannot resist any more than your arm can resist when you want it to do something."[37]

The inaudible or "silent" treatment often gave patients and critics the mistaken sense that nothing was happening, but by thought alone the healer would be arguing (reminding herself or the patient) that sickness, sin, and death cannot exist because God is All and God is Good. In the Nebraska trial of Buswell the state's attorney hammered this point to determine if there had been any evidence that Buswell had done anything at all when he treated patients. A. Parker, whose nine-year-old daughter had died of diphtheria on 21 January 1893 while exclusively under Buswell's care, testified that Buswell did not give or prescribe any medicine during his daughter's illness but prayed with her at least daily during the week before her death. The state's attorney asked:

Q. Did you ever hear Mr. Buswell pray?
A. No sir.
Q. You didn't hear him pray?
A. No sir.
Q. How do you know he did pray?
A. They believe in silent prayer. . . .
Q. What did he do?
A. Sit down by the bed side, . . .
RECROSS EXAMINATION, By Mr. Hazlett [Buswell's attorney].
Q. Tell the jury what attitude Mr. Buswell was in when he was at the side of your daughter's bed?
A. He was holding his face in his hands, seemed to be in prayer.[38]

During the 1870s and 1880s Christian Science healing seemed, to many outside observers, to be nothing more than faith healing. Some Christian Scientists, therefore, eager to assert a distinctive identity, argued that the efficacy of their silent treatment decisively proved their uniqueness. Unlike faith healing, in which the speed and completeness of a healing was directly proportional to a sick person's faith in God or the apparently capricious benevolence of God, Christian Science inaudible treatments proved that healing depended solely on a practitioner's understanding of Science and not on a patient's faith.[39]

Not all Scientists, however, shared this view of silent treatment as an acid test. Others believed that a patient must have at least some confidence in Science for the practitioner to be effective.[40] And still others took a middle position, arguing like Willard S. Mattox, practitioner and assistant to the manager of the Committee on Publication, that "there are two essentials to a Christian Science treatment. The practitioner must know enough to heal, and the patient must know enough to be healed."[41] Eddy took a similarly balanced position on the issue. On the one hand she asserted that patients did not have to exhibit a mature or complete faith in Christian Science, but only enough faith to give it a try. Patients had experienced dramatic healings, she believed, even though they had only "tried it . . . because their friends wished them to."[42] On the other hand, she agreed that even if a patient lost understanding and used material remedies, "the practitioner must go on, nothing doubting, nothing fearing, to the very foundation of the malady, over the waves of mortal mind to the sin which was its source, and cleanse his patient of that. The causing sin made void, is the body made whole *in every case*."[43] Nonetheless, as early twentieth-century Scientists came increasingly to understand healing as a process of growth and enlightenment rather than a sudden "miracle," they began to emphasize patients' individual contributions to their own healing.

As the patient became more familiar with the basic principles and peculiar language of Christian Science, the healer became less reluctant to speak audibly. Richard Walthers, a later witness at the Buswell trial, testified that Buswell was treating his fourteen-year-old nephew for a running sore on his leg "according to the Christian Science doctrine." Defense attorney Hazlett asked Walthers if he had seen "Mr. Buswell talk to the boy or to you or to any of you concerning the boy's sickness, have you heard Mr. Buswell talk to the boy, if so what did he say?" Walthers replied that Buswell "told him to trust in the Divine power and that God would heal him and that he should not get angry but hold peace and love."[44] New York City practitioner Willis V. Cole described his treatment of a female "patient" (an undercover County Medical Society investigator posing as a patient) in the following terms:

She sat about four feet away from me, facing me. She told me that she had come to be treated for trouble with her eyes and stomach trouble. I informed her that Christian Science treatment was prayer to God, we did not believe in Drugs, medical treatment, anything like that, and she asked me to give her treatment. Something was said in regard to the basis of

Christian Science, and I told her substantially that Christian Science was the truth about God, and the truth about man and the truth about man's relationship to God and the truth of his birthright as a result of this relationship, which is the foundation of what we teach, and I told her that on this basis disease was no part of her birthright, or inharmony, and when she realized her oneness with God, and got in harmony with God that this was the treatment and was what we would do. She sat there for about fifteen minutes. I covered my face with my hands, or sat with my head partially bowed for fifteen minutes. . . .

Q. In prayer?

A. In prayer to God, nothing more, praying to the Almighty.[45]

When the practitioner spoke or wrote to her patients, she reminded them of the principles of Christian Science doctrine, recommended passages in the Bible or Christian Science literature (especially *Science and Health*) for regular or specific reflection, and specifically argued against their belief in particular symptoms. The suggested readings often included sections that correlated with a patient's symptoms. For example, in 1899 Irving C. Tomlinson of Concord, New Hampshire, read a passage from *Science and Health* to Jennie A. Spead, who was suffering from severe stomach pain due to appendicitis and said, referring to the passage, "Here is a mustard plaster for you."[46] Similarly, Thompson counseled one patient suffering from a swelling to "try and know that 'Love is not puffed up.' The seemi[ng] testimony of the senses, and the material[i]ties apparen[t] all about us are not true, but GOD is true and He it is that is all around us and nothing else."[47]

The practitioner also encouraged her patient to give up treatments or regimens prescribed by a regular physician. Thompson once advised a patient, "You are spiritual, you reflect LOVE for that is all the[re] is to reflect, you can eat anything you desire to without any inconvenience to you."[48] A witness at the 1911 trial of Willis V. Cole testified to the following recommendations given by Cole:

Q. When you told him you had a pain in your back, what did he say?

A. I then says that I had a porous plaster on my back at that time; and I said to him what did he think about the pain I had in my back. He said it was some kind of disease but he could not tell what it was. He said "I can cure it."

Q. Did you say anything about this porous plaster?

A. I said I had a porous plaster on my back with the pain that was there.

Q. What did he say to that?

A. He said "You must now take off that porous plaster because Christian Science cannot cure with plasters on."

The witness also wore eyeglasses, and when the attorney asked the witness what the defendant had said about that, he replied:

A. This defendant said that I should have more faith and understanding, that I must have more courage, that I should remove the glasses. . . .

Q. What did he say when you told him that—that you would have to keep them? Did he say anything?

A. He said if I wanted to be cured by Christian Science, I must remove the glasses.[49]

If a practitioner's understanding of Science focused on the importance of her own recognition that all is good, then she treated children the same as anyone else. The following short piece of instruction by Eddy advocated such an approach: "If our children need treatment, do not sit up all night and treat, but treat yourself, and go to bed and to sleep."[50] If, however, the practitioner believed that a patient must become clear in his own mind regarding the true nature of reality, then the treatment of children, and especially infants, constituted a special case because of a child's limited cognitive powers. According to Buswell,

we [Christian Scientists] think the infant is influenced largely by the thoughts of its parents. To illustrate, sometime ago I was at a home where a little infant seemed to be suffering from a fever; the father had tried to help the child but it, seemed not to improve. He asked me for help at the supper table; he said I have been telling my wife that all that ails the child is her fears and the mother seemed to think that that was the case and she was bowed down in grief. I told her that there was but one power and that was love and there was no power in her that could heal her child and when she was convinced of that fact it seemed as though a light had broken in upon the mother and she was [c]heered by it and the condition of the child responded in harmony and love [it recovered].[51]

In their treatments practitioners avoided memorized formulas, even for cases that seemed very similar, because they wanted to remain open to the leadings of truth for each patient and avoid the appearance that healing could be mechanically invoked by magical chants. Although Eddy told her students that "the sick deserve our sympathy and should claim our efforts more even than the sinner," some practitioners still feared that a display of sympathy might only strengthen a patient's symptoms or that they themselves might acquire the patient's symptoms.[52] Therefore in 1903 a practitioner warned readers of the *Christian Science Journal* that "the time has passed when the tyro in Christian Science can feel justified in answering tales of suffering and grief with the assertion, 'There is no such thing as pain, and there is nothing in the world to grieve about.'"[53] To this day Scientists feel it necessary to deny the charge of insensitivity, arguing that they exhibit *true* compassion when they lift people out of trouble by opening truth to them, not by sharing in their distress.

Undoubtedly, patients a century ago appreciated the individualized care of practitioners, which often contrasted with their experiences with physicians. Neophyte practitioners, however, often repeated the experience of one veteran healer who noted that "for a long time after I began to practise, the unreal was so much more realized than the real, that certain, careful steps in scientific argument (called formulated treatments) were necessary to bring me into that state of consciousness which recognizes that man *is* whole."[54] Also, as a practitioner settled into a healing routine, she could see that some types of patients or patients with certain complaints seemed to respond to similar kinds of "arguments." Practitioners often recommended a regimen of regular and persistent study of key passages in *Science and Health* that, when memorized, could be recalled rapidly during times of stress or discomfort.

In an effort to systematize such treatments, many generic Christian Scientists (and no doubt some orthodox Scientists as well) followed and published numerous step-by-step formulas for healing. Despite her own formulaic healing instructions in early editions of *Science and Health* and *The Science of Man*, Eddy later altered her views and warned her followers to avoid memorized formulas drawn from their own experiences or her writings. She denounced the use of repeated affirmations of truth or denials of errors as healing incantations and accused "schismatics" like Luther M. Marston and Ursula N. Gestefeld, who persisted in publishing their own summaries of Christian Science, with trying to confuse the clear truths she had discovered. Henceforth orthodox practitioners were to give individualized treatment to each patient.

"Compensation": Fees

Before 1910 practitioners usually charged the going rate of physicians, $2 for the first visit and $1 for each subsequent treatment. For the extended treatments involved when a problem failed to respond quickly, they reduced their charges, often charging only $5 a week. In January 1910 Eddy directed that "Christian Science practitioners should make their charges for treatment equal to those of reputable physicians in their respective localities."[55] In response, practitioners doubled their prices to $4 for the first visit and $2 for each following treatment and raised their rates for absent treatment to $7 a week.[56] If Thompson was typical in encouraging his absent patients to visit his office twice a week for treatments when possible, at an additional cost of $1 per visit, then the base price for a week of treatments could have reached $9.[57] Subsequent price hikes reflected the rising fees charged by physicians, but practitioners, like M.D.s, continued their practice of reducing their charges for people in poor financial circumstances. The following is a monthly statement Thompson sent to a patient who had unidentified but serious problems:[58]

Jan. 10. 1911.

Mr. A. F. Adams,
      37 Park Avenue Newton, Mass.
      To John H. Thompson, C.S.B.

Dr. [debit]

For professional services 35 days absent treatment from December 1.1910 to January 4, 1911.	$ 35.00.
For 28 professional visits during same time	$112.00.
For extra help 2 nights	$ 2.00.
	$149.00.

Evidence regarding annual incomes is scarce, and much depended on whether a healer worked part time or full time. James Neal, recalling his practice in Kearney, Nebraska, stated, "My practice during that six months [ca. 1888], while very much of it was charity, netted me a little better than $3500." Two practitioners, one practicing in Chicago during the 1880s and the other in New York during the early 1900s, each reputedly grossed between $5,000 and $6,000 a year, and that may not have been uncommon for large urban practices.[59] Doubtless more typical for rural areas and fledgling practices was Thompson's experience when he first started his practice; he could not quite make ends meet for himself, his invalid wife, and their daughter on an average income of $50 a month.[60] Buswell kept no account books, but he estimated that during the eighteen months before his trial in February and March 1893 he had treated "a hundred or more" persons around Beatrice, Nebraska, and in the past had received for his assistance of the sick "not more than one hundred dollars a year. Perhaps not near that, a very small amount at any rate."[61] An 1895 survey taken by the editors of the *National Medical Review* put the average income of all the regular physicians in Washington, D.C., at $2,000; however, in the 1890s the Washington Pension Bureau hired newly graduated male physicians at annual salaries that ranged from only $1,200 to $1,800.[62] If, as historian Ronald L. Numbers has established, "well into the twentieth century, most American physicians earned less than $2,000 a year," then comparably speaking, at least some practitioners did quite well for themselves.[63] Before the movement censured specialization in the early twentieth century, practitioners who developed a specialty—in "dentistry" (especially the use of Science to control pain), "obstetrics" (the use of Science to control the "illusions" of childbirth, that is, belief in the reality of anatomy, physiology, physical intercourse, and pain), or absent healing—held a competitive edge.[64]

Frequently critics reproached practitioners for charging patients when Jesus and his disciples had given freely of the truths that Scientists now claimed to have recovered. Practitioners retorted that every worker is due a just recompense for her costs, her labor, and her time. In the words of the editor of the *Christian Science Journal*, "The minister is undisputedly entitled to remuneration. The physician, who claims his remedies are provided by God, makes his charges and receives his pay without question. If their premises are well-founded, upon what principle can they deny the right of Christian Scientists to fair remuneration?"[65]

Maine practitioner George W. Wheeler went so far as to file an action of *assumpsit* against the estate of Andrew M. Oliver, a former patient who had died intestate in 1888. Wheeler had treated Oliver for ten weeks at a charge of $3 per week. For some time Wheeler had tried to collect on the debt, but to no avail. Prince A. Sawyer, administrator of Oliver's estate, argued that "so-called 'Christian Science' is a delusion,—that its principles and methods are absurd, that its professors are charlatans— that no patient can possibly be benefitted by their treatment." The Supreme Judicial Court of Maine, however, concluded that Oliver had chosen "that treatment, and received it, and promised to pay for it. There is nothing unlawful or immoral in such a contract. Its wisdom or folly is for the parties, not the Court, to determine." It found judgment for Wheeler in the amount of $30 with interest.[66]

Although technically the payment was for the *time* the practitioner spent and not for the *cure*, practitioners did not always make the distinction explicit to their patients, and doubtless many patients, had they considered it, would have viewed it as a distinction without a difference.[67] At the 1893 trial of Buswell, former patient C. N. Bennett insisted that neither Buswell nor his wife made charges or kept an account book. Rather, he understood that "if I wanted to pay any thing it was all right."[68] Another witness, Peter Burgis, had even tried to get Buswell to "set some price that he ought to have," but Buswell responded that "he never set any price[;] if I thought I was benefited I could donate him any thing I pleased."[69] However, the testimony of a subsequent witness, L. Bushnell, made it clear that Buswell expected payment: "He told me once how much he paid out of his own money and that he would like to get that back and . . . that if I was not able to pay him any thing more than that [it was] all right."[70] Buswell himself stated that "as a rule" he did not charge people when they came to him for "advice"; instead, "we tell them that we leave the question with them and God," but "Jesus says the laborer is worthy of his meat and we expect that those who we spend our time for [t]o remunerate us for it."[71]

Buswell may not have formally billed his patients, but many practitioners did, and they also made it clear that the responsibility to pay was not simply a question of fulfilling one's moral obligation to pay debts. At the least, payment allowed practitioners to pay their own bills. In 1910 Thompson wrote to a patient:

> In sending out my monthly statements I notice your bill still unpaid and trust you will find it convenient to pay it by this time; I noticed what you said in your last letter about Christian Scientists being anxious for their money, but of course you and everybody admit they cannot pay their debts

without it, if I were to buy farm produce from you[,] you would certainly expect me to pay for it, if money had not been in some way contributed to Jesus he could not have carried on his work.[72]

But more important, patients' willingness to pay gave evidence of their confidence that healing had already occurred, which in turn sped up the healing process. One practitioner reported in 1885 that she had "noticed in practice that selfish people, dishonest people, stingy people, are 'slow cases.'" As a young practitioner, she had accepted patients "who came as a last resort but did not want to pay." Nevertheless, she had "worked faithfully for them. But I never cured one of them. Why? 'There is no remission of sins without sacrifice.'"[73]

"Demonstration": Healing

When a patient experienced healing, Scientists asserted that the practitioner and patient had "demonstrated" over the affliction. The word "demonstrate" captured both the confidence that a metaphysical argument had awakened one to the true nature of reality and the sensory demonstration that symptoms had disappeared. Treatments did not so much destroy disease as remedy an ignorance of health.[74] Although Christian Scientists resisted similar comparisons by turn of the century contemporaries, Christian Science healing experiences were not unlike those historian Grant Wacker has ascribed to the followers of John Alexander Dowie (1847–1907), founder of the Christian Catholic Apostolic Church in Zion. For Christian Scientists, as for Dowie's followers, "healing represented not only a release from pain but also a sacrament, a palpable symbol, of those rare but unforgettable moments of grace in the life of the believer."[75]

Sometimes healing was slow and sometimes it was rapid, but according to Christian Scientists it never failed, except in the sense that the practitioner or her patient had not yet gained a complete understanding of the true goodness of reality. In the pages of the *Christian Science Journal* practitioner Herbert S. Fuller decried the way testimonies to quick healing dominated testimony meetings while stories of slow healing were "not told because error says that they are not wonderful enough, when the fact is that the simplest demonstration is wonderful."[76] The early twentieth-century struggles of Mr. and Mrs. Oliver W. Marble in

Sandusky, Ohio, illustrate the poignant and often conflicted experiences of Christian Scientists as they struggled to live up to their ideals in the face of sensory experiences not yet "overcome." But their experience also typified the tenacity of belief that often resulted when Scientists personally experienced a dramatic "demonstration." During Marble's 1903 trial for practicing medicine without a license, the prosecutor, Roy H. Williams, led the accused practitioner to admit that he had in the past sought the services of a physician for the treatment of his son's epilepsy. " 'Then your faith in Christian Science failed?' said Mr. Williams. 'No,' replied the witness. I felt that my progress in Christian Science was not sufficient to meet the demands of the case.' " Choking back tears, Marble then proceeded to tell the whole story. The boy had been an epileptic since his sixth birthday, and his parents had gone from physician to physician seeking a cure until finally the physicians at the state hospital for epileptics "pronounced his case a hopeless one." The parents turned solely to Christian Science therapy early in 1903, when Andrea Dosser Proudfoot, C.S., a friend in Chicago, invited the child into her home for treatment. Six weeks passed with no change, but then

> one day in February he suffered three severe spasms. The lady discovered the boy in his room and at once invoked the aid of Christian Science when the pain left the boy and he was cured. The permanence of the cure has been evident from the fact that the child has not suffered any recurrence of the symptoms but on the contrary has grown both mentally and physically. Tears of gratitude coursed down the cheeks of the witness as he concluded his story by saying: "I thank God for Christian Science."[77]

The Marbles could have taken some comfort from the advice found in the June 1892 *Christian Science Journal:* "Why not admit the truth? that, while a knowledge of Christian Science enables one more easily to prevent, lessen and overcome the ills of life, there is no Scientist who is wholly exempt, at all times, from aches and pains, or from trials of some kind."[78]

Many Scientists agreed that patients experienced faster healings when they quickly grasped a clear understanding of Science. Practitioners who could stimulate speedy and long-lasting results were in great demand and earned the envy of their colleagues. On occasion Eddy reprimanded highly successful practitioners and tried to restrain their activities. In early 1889 the *Christian Science Journal* adopted a policy not to publish practitioner names next to healing testimonies, because in the past "reports of cases have sometimes been too much in the nature of certificates to a supposed personal power of the healer, instead of testimonies to the power of Truth."[79] The effect of the policy is unknown, but for those who disliked it, the market provided ample room for membership in divergent schools of Christian Science or mental healing.

How did patients describe their healing? Most of our knowledge of patients' experience has come through the often stylized testimonials to healing that played such an important part in strengthening the faith of Christian Scientist believers and stimulating the interest of potential converts. However, the historical record provides brief, tantalizing glimpses of what a patient experienced during healing. Many reported feelings of assurance or peace and a sense of renewal, but James Elerbeck vividly described a powerful psychophysical experience after being treated by Buswell for a rattlesnake bite.[80] Buswell discussed some biblical passages with him, then Elerbeck "lay down on the lounge and he set down and put his hands over his face and was in that position may be ten or fifteen minutes and *all at once I felt it come right through me and it raised me up* and I sit on the lounge and I told him I had waken up and from that time on I had no more pain only there was one or two minutes when I first got up and put my feet to the floor that the stiffness seemed to be hard, for a few minutes."[81]

"Infancy of Truth": Limitations and Failures

Despite the optimistic cure rates Scientists claimed for all sorts of physical and mental conditions, Eddy herself conceded in the first edition of *Science and Health* (1875) that "the present infancy of this Truth so new to the world" demanded that Christian Scientists should "act consistent with its small foothold on the mind." She encouraged her followers to consult surgeons for bonesetting but then to "let science facilitate the knitting process, and re-construct the body without pain or inflammation as much as possible in these days of ignorance."[82] This, in effect, avoided the difficulty of denying the senses when confronted by a serious fracture; in part for similar reasons she later granted practitioners permission to seek a physician's attendance at childbirth, while they treated the patients' errors of belief in anatomy, pain, and anxiety. Consistent with these allowances, practitioner John R. Hatten of Dayton, Ohio, explained at his

1895 trial for manslaughter that in treating broken bones he "treated pain. The limb would be set. He would not use morphine to deaden pain but resort to prayer. He believed that prayer is more efficient than morphine."[83] Such exceptions to "spiritual law," however, made it awkward for practitioners to appear consistent, especially on a witness stand. C. A. Clark, first reader of Milwaukee's Church of Christ, Scientist, and a practitioner for many years, had a hard time in 1900 explaining why he could not heal broken bones but supposedly could heal tubercular lung tissue. Milwaukee's assistant district attorney asked incredulously:

Q. . . . But you claim that although you can not set bones so as to heal them you can restore lung tissues in cases of consumption by mere understanding?
A. It is fortunately done so that there is no doubt about it of any kind or shape or manner.
Q. Have you ever treated and set fractured bones by your method?
A. I haven't had any cases recently.
Q. Do you know of any cases that have been treated?
A. I have read of them.
Q. But still Mrs. Eddy herself suggests and says that it is better to call a surgeon in the case of fractured bones?
A. Yes.[84]

Consulting with physicians seemed to vary from practitioner to practitioner. Many refused to handle a case, in the words of Emma S. Davis, "with doctors and medicine in the house."[85] But with surgery permitted for broken bones, what other grossly obvious anatomical situations, such as swelling or mutilation, might surgeons treat? In California, San Bernardino County District Attorney T. J. Fording put the issue clearly to Christian Science practitioner Emma Davis in her 1892 trial:

Q. Does Christian Science accept either medicine or surgery?
A. It does not accept medicine.
Q. Does it accept surgery?
A. In case of a broken bone it accepts any surgery. . . .
Q. Is it the duty of persons who profess and practice Christian Science to acquaint themselves with the human anatomy?
A. Not at all.

Q. Do they attempt to acquaint themselves with the laws of hygiene?

A. Does the Christian Scientist?

Q. Yes, madam.

A. No, sir.

Q. Do they pay any attention to physiology?

A. No, sir.

Next, after describing in detail the progressive development of symptoms—swelling, protuberance, delirium—suffered by the man whose death had provoked this trial, Fording asked:

[Q.] [W]ould you regard that [as] a case in which surgery was demanded? . . .

A. In such a case as that I should consult the family, [and] the nearest friends of the patient. I should ask them if there was any indication of fear that their friend was getting worse. I should ask them if there was anything else that they wanted to do, if I had the least apprehension that they were not satisfied, that they had any fear in connection with the treatment, I should most assuredly advise them to have it. That is what I would do in that case.

Q. In a case like that which I have named, when the swelling continued to enlarge day after day, would it be the part of a person practicing Christian Science to continue to send out words to friends that the patient was doing well, and that they should have no fear?

A. I must answer all the questions according to my own understanding, as if I was in the case?

Q. Would it be a proper act on the part of a person practicing, as you understand the science?

A. If a Christian Scientist knew that a patient was growing worse—if they knew they were growing worse, it would not be the right thing to do to send word out that they were get[t]ing better.[86]

Fording believed he had elicited a response establishing that Davis believed a competent practitioner would acknowledge obvious physical symptoms such as severe and progressive swelling. As we have seen, practitioners, especially during "diagnosis," did not completely ignore their physical senses, so Fording was possibly correct. But if so, he was only partly correct, for Davis also meant that a practitioner must report what she knows to be true—she must be honest. The understanding of what is true—God is good and disease does not exist—and its honest affirmation are what heal. For a practitioner simply to affirm something she did not herself believe to be so would be tantamount to malpractice.

During the last decade of Eddy's life, laws regulating the practice of medicine drove Christian Scientists to accommodate doctors and their medicine, and Eddy's own experiences hastened the journey. Eddy instructed parents to submit their children to compulsory vaccination and urged practitioners to comply with laws requiring the reporting of contagious diseases. After occasionally using morphine to deaden the pain of her own kidney stones, she sanctioned the use of painkillers for bouts of intense pain, even though she had claimed in 1884 that medicine "only weaken[s] your power to heal through mind."[87] Keeping to the spirit if not the letter of these concessions, Thompson advised a patient in 1909 that "it is admiss[i]ble to use a disinfectant" on an infected foot, but that to use it "with a sense that it has any curative quality whatever would only hinder the demonstration by divorcing your mind from the true source of healing and diverting your thoughts from what is true." However, even such adaptations had bounds. Responding to the patient's further question about using regular physicians, Thompson asserted that "to have those who are in sympathy with material methods of healing and so opposed to spiritual methods examine and superintend the work being done by the Mind practitioner only retards the work and gives that Practitioner much extra work" and warned that "I owe it to Christian Science and myself to withdraw from this case if the material Doctor is to be consulted."[88]

Christian Scientists believed that death was as much an illusion as sickness and predicted that as society became more familiar with the principles of Science, demonstrations over death itself would become more frequent and death rates would decline. Many Scientists believed, as did Eddy herself, that several times her healings raised the dead.[89] Nonetheless, the tenacity with which this error of belief in death attached itself to human beings made it one of the most shocking claims to confront and often the most difficult to treat. How did a claim of death present itself to a practitioner? Although Christian Scientists no doubt experienced death in different ways, Emma Nichols, a practitioner in Milwaukee, Wisconsin, for eleven years, recalled her encounter with a young patient's death claim in 1900 in the following way:

I took off my wraps and sat down at the foot of the bed and gave the child a treatment. A few moments afterwards I thought—the thought came to me that this was a claim of death. . . . I thought when I entered the room that it [her breath] was rather quick, but in a few moments that thought did not come to me, because another thought came; there seemed to be a voice— seemed to be with the breathing, and that, as she breathed it seemed to me unnaturally long and deep, and that was what suggested the thought of death.[90]

After a seven-year-old girl died of diphtheria in 1902, John L. Roberts, second reader in New York City's Fifth Church of Christ, Scientist, reported to the press that "the Quimby child is not dead. Death is simply a transition from one state to another. The child is now living in another state and is perfectly happy."[91] Little Esther's mother testified during the subsequent grand jury hearings "that after the child had 'passed on' she sought to recall her to life by the 'life thought,' but the thought would not work, little Esther had 'passed on' too far."[92] John R. Hatten of Dayton, Ohio, testified at his 1895 trial for manslaughter that there had been "considerable excitement" when a physician, summoned to sign a death certificate, declared the child dead. When the coroner asked "Why [the] excitement?" Hatten replied, "We are human."[93]

Even when death seemed to signal failure and foreclose the possibility of any future "argument," the practitioner affirmed her belief in the unreality of matter by denying the reality of death. Borrowing novelist John Buchan's metaphor for fishing, we might say that the Christian Science practitioner provided for her patients "a perpetual series of occasions for hope" that extended even to the grave and beyond.[94]

Separating "True" Scientists
from "Pseudo" Scientists

A singular phenomenon is just now observable. The market for Science is
going up, but the appreciation of the individual Scientist is going down.
Truth is working its way into the general consciousness, but the true and the
pseudo Scientists are lumped together, and a general average on the lot is
struck by the uninformed, but disinterested and honest public. . . . To dispel
the confusion in the public mind as to who are and who are not Scientists, we
must first get rid of it in our own minds.
　　　　　　　　　　　—*Christian Science Journal*, 1889

Abby H. Corner, a fifty-year-old woman "with soft light hair sprinkled
with gray, penetrating blue eyes, a remarkably pretty and intelligent face,
[and] a robust, yet well-formed figure," had been a Christian Science
practitioner for about three years, although she and her husband had remained members of their local Methodist church.[1] She had studied Christian Science with Mrs. Dr. J. C. Cross and subsequently had graduated
from Eddy's Massachusetts Metaphysical College. At about eight o'clock
on the evening of 19 April 1888 Corner's daughter Lottie A. James of the
Boston suburb of West Medford, already "quite sick from vomiting,"
went into premature labor.[2] Taken by surprise, the household soon found
itself grappling with a situation that seemed to worsen at every turn.

　　Corner called Henrietta Hegring, a nurse-midwife "opposed to
'Christian science' principles," away from a local Methodist prayer meet-
ing to attend her daughter, while Corner herself applied the techniques of
Christian Science healing to her daughter's "claim" of childbirth. During
James's "incessant moaning [her mother] frequently interposed, 'You are

not suffering, Lottie; just be brave and put your trust in God and our prayers.'" At about 10:30 P.M. she gave birth to a stillborn daughter whom the nurse-midwife treated just as she had "dozens of other children—washed it and sought to bring it to life—but to no avail. The nurse then turned her attention to James, who was in great pain and suffering severely from hemorrhage." James pled with the women "to send for a physician, saying: 'Oh, mother, if you knew how I am suffering, you would send for a doctor.'" Near eleven o'clock James experienced convulsions that spurred her mother to send someone for another Christian Science practitioner and the nurse to request the services of Dr. J. L. Coffin; however, before Coffin had traveled the half mile from his home, "mother and child were dead." Believing that the case merited investigation by public officials, Dr. Coffin told Corner "not to disturb the body in any way." Corner signed a death certificate that listed the cause of death as "childbirth after three hours' illness," but a subsequent autopsy revealed that James had bled to death.

For days feelings of grief, anger, incredulity, and recrimination kept the untimely deaths and the Christian Science practices that surrounded them the topics of conversation "in the corner groceries, the drawing rooms, and every place of public gathering" throughout the Boston area. According to town physician J. E. Clark, who observed the autopsy, "the mother and child were both in a perfectly normal condition until within an hour or two of death, and . . . with the care which could have been given by any midwife or nurse, the birth would have been natural and free from danger." The district's medical examiner, Dr. Thomas M. Durrell, asserted, "If ever there was A Case of Manslaughter this is one. Had the woman been struck over the head with a club it would not make a stronger case than this." Dr. Coffin agreed, but he doubted that any case could be made against Corner given the "loose laws of this State in regard to the practice of medicine," which allow "quacks, knaves, fools and idiots" to practice medicine. The *Boston Daily Globe* threw aside all semblance of impartiality to report that "the prayer test has been tried in West Medford, and this time with fatal results, two lives being sacrificed to a superstition that seems almost criminal. . . . A faith so ineffectual that it will stand calmly by while an innocent mother and an unborn baby die, needs correction if not extermination." Abby Corner herself reportedly attributed her daughter's death to "an inscrutable decree of Providence," consoled herself with the belief that "He doeth all things well," and bore witness to her beliefs by asserting that she was "glad that drugs at least did not kill" her daughter.

Faith curists, mental healers, pneumatopaths, and Christian Scientists all lined up to publicly distance themselves from the unfortunate Corner. Boston's Charles Cullis, M.D., well known for his belief in the use of Christian faith healing to cure diseases that medicine could not, disavowed any system that substituted prayer for medical care. In his view faith cure and Christian Science were "as different as Buddhism and Christianity," and the people running the Christian Science colleges were just in it for the money.[3] Luther M. Marston, M.D., president of one of those colleges (the Boston College of Metaphysical Science) and a former student of Eddy's, admitted that "many of the students and professors make mistakes and attempt too much." He further ventured that "in such a case as that at Medford perhaps it would be better for the metaphysic[i]an to call in a regular practitioner." Even her old teacher, Mary Baker Eddy, abandoned Corner to an incensed Boston. In an open letter to the *Boston Daily Globe* and the *Boston Herald*, Eddy wrote that this "lamentable case . . . should forever put a stop to quackery." "All professions are subject to impostors, Christian Science included," she continued, "but the history of science is by no means at the mercy of charlatanism. Recreant practitioners in any school of medicine are a disgrace to it."[4]

Moving beyond the specifics of the Corner case, Boston-area physicians and lawyers debated the pros and cons of legislation to regulate the practice of medicine. Dr. O. S. Sanders believed that the state legislature's recent failure to establish a board of medical examiners proved they "do not know what they are talking about." But while they dithered, "Boston is loaded down with presuming medical characters. They come here, I might say, from everywhere. They put out a sign and call themselves doctors. A stranger coming to Boston, if he is taken sick, is just as likely to visit a quack as a regular graduate. In case he does visit the quack, he generally gets taken in. This does not speak very well for the city of culture, does it?"

Homeopathic physician Conrad Wesselhoeft agreed that the legislature should look into the matter of licensing, but he feared it would "be some time before the differences between rival schools of medicine will permit them to come to an understanding as to how the matter can best be regulated." Anyway, he added, "you can't legislate fools into wise men.

If a man prefers to be killed by a quack rather than cured by a regular physician, that's his affair, and the law should not interfere with individual rights."

Such was the highly charged atmosphere on 24 April 1888, when Abby H. Corner was arraigned before Judge John W. Pettingill's First District Court of Eastern Middlesex County to answer the charge of manslaughter. Pettingill set bail at $5,000 and continued the case until 5 May. When the court finally reconvened on 18 May, the case was no longer front-page news, and when Pettingill found Corner "probably guilty of manslaughter in causing the death of her daughter" and ordered her held for trial in the superior court at Cambridge (the case never came to trial), the *Boston Daily Globe* reported it on page 8.

Before these events of 1888 there had been at least three trials of Christian Scientists charged with practicing medicine without a license; however, Abby Corner was the first practitioner charged with a felony (see appendix). The gravity of the manslaughter charges contributed to the furor that surrounded the case, focused public attention on unorthodox healing in general and mental healing in particular, and heightened the importance of considering the government's proper role in regulating healers. But the Corner case did more than agitate the general public and precipitate a public relations disaster for mental healers and Christian Scientists. It crystallized a debate that had been developing for some time among Christian Scientists themselves over the behavior of Mary Baker Eddy and the nature of her authority. More than anything else, the side Christian Scientists chose in that debate marked them as "true" or "pseudo" Scientists.

Orthodoxy, Heterodoxy, and Heresy

In their efforts to define orthodoxy, Christian Scientists wrangled over nuances of healing theory and practice, feuded over the documentary and historical origins of their movement, and bickered about a Scientist's proper relationship to Christian denominations; but they especially fought and fragmented over how far any leader, especially Eddy, should be allowed to exert authority over Christian Scientists. Throughout their history Christian Scientists, like many of their fellow nineteenth-century American sectarians, struggled to create and then preserve their identity by claiming to possess "new light" that radically transformed human experience. Yet rarely if ever did such "new light" arrive with a fully developed and unchallenged authority. Rather, it had to withstand scrutiny and undergo severe tests before a community confidently used it to establish boundaries between orthodox and heretical teachings or behaviors, between true and false teachers, or in the case of Scientists, between authorized and unauthorized practitioners.

When Christian Scientists scrutinized Eddy's claim to be the "discoverer" of "new light," they struggled to interweave the divine and human dimensions of prophetic authority. A portion of a person's self-identity as a religious prophet comes from her belief that she has encountered God or his revelations in a way that radically transforms her vision of the world. A "prophet," in the words of biblical scholar and theologian Abraham J. Heschel, "is a man who feels fiercely. God has thrust a burden upon his soul, and he is bowed and stunned at man's fierce greed." The prophecy that results "is a form of living, a crossing point of God and man. God is raging in the prophet's words."[5] But accompanying the prophet's belief that God has chosen her to bridge the gap between the sacred and the profane is the recognition that, initially at least, the radical nature of her message will make her mission difficult and her followers few. Undoubtedly the prophet's interaction with her own culture molds the messages she "hears" from God's sacred realm, but her culture also reinforces her prophetic self-identity by the way society hears, obeys, or rejects her message. Filled with a sense of responsibility to share her vision with the world, she feels discouraged by the world's resistance to her message but reinvigorated and self-confident when followers gather. In either case the prophet's audience has helped to authenticate her call: the worldly (often the majority) by rejecting the divine word and refusing the narrow path of truth, and the saints (often the minority) by heeding the prophetic injunctions. As a result of this interaction between prophet and audience, the vision she receives from God and the way she views her world change as the prophet, inextricably tied to her social context, strives to fulfill her mission.

Eddy's experience followed such a pattern. Although open to both the call of God and the influence of her culture, she remained especially attentive to her contemporaries' remedies for her physical and religious shortcomings and always implemented natural means to ensure the survival of her movement. A prophet's authority, however, depends on a social identity as well as a self-identity; prophetic charisma is, as sociologist Bryan R. Wilson has argued, "a social phenomenon, not a psychological personality type."[6] The ability of a prophet and her followers to mold their beliefs and practices and to exact obedience is the final test of prophetic authority. Eddy's own authority, although shaky at times, grew as her followers acknowledged the centrality of her revelations to their movement's identity, pruned away those dissenters who refused to accept the new image, and adhered as "true believers" to the new commandments that poured from her pen.

The defection in Lynn, Massachusetts, of some of Eddy's key students during October 1881 because she had made manifest her departure from Christ's way by "frequent ebullitions of temper, love of money, and the appearance of hypocrisy" signaled the beginning of a long struggle to define the nature and extent of her authority to distinguish the "true" from the "pseudo" Christian Scientists. Stunned by the apostasy, Eddy received solace through a vision witnessed by a handful of her faithful followers, who jotted down the disconnected but divine messages of woe and exhortation. Strengthened by this sign of divine favor, Eddy's faithful flock ordained her pastor of the Church of Christ, Scientist, and publicly asserted that "she has had little or no help, except from God, in the introduction to this age of materiality of her book, *Science and Health*" and that "we do understand her to be the chosen messenger of God to bear his truth to the nations, and unless we hear 'Her Voice,' we do not hear 'His Voice.'"[7] Underscoring the new beginnings, Eddy shook Lynn's dust from her feet and moved her headquarters to Boston.

On arriving in Boston, Eddy and her faithful few planted and nurtured the institutions they believed would bring true healing to all the nations of the earth. The Massachusetts Metaphysical College opened in 1882, and its first class graduated in May. In April 1883 Eddy edited the first issue of the *Journal of Christian Science* (renamed the *Christian Science Journal* in 1885), a monthly publication intended to encourage her students and to protect them from the influences of other mental healers. An advertisement for the *Journal* that appeared in the 1884 bulletin of the Massachusetts Metaphysical College touted its role in health and healing: "AN independent family paper to promote health and morals. Gives the practical work of Christian Science. All invalids should have it. Price, $1.00 per year."[8] To further enhance the spread of truth and ensure orthodoxy of belief, during August 1884 Eddy taught the first Normal class for the training of authorized teachers, whom she encouraged to open "colleges" and establish associations of their own. And in response to their leader's suggestion, in 1886 her followers formed a National Christian Scientist Association, whose annual meetings gave Eddy a platform for exhorting and unifying the graduates of her college and its sister institutes.

But beyond offering a fresh start, Boston presented new difficulties, as she soon discovered. During the 1880s New England's cultural center swarmed with physicians and preachers of all kinds and boasted a well-established and highly competitive population of mental healers as well. Although such an atmosphere provided fertile ground for Eddy's new seeds of truth, it also presented ample opportunity for the enemies of Truth to sow their own weeds of falsehood and discontent. Many American mental healers first learned of mind healing by hearing about Eddy, by reading *Science and Health*, or by taking classes from the self-styled "discoverer" of Christian Science herself, but they often remained ignorant of or uninterested in any claims Eddy made for priority or authority over doctrine. They simply lumped her publications with those of other mind healers who supplied healing advice and envisioned a benevolent universe. Unaware of or unconcerned with the Lynn declaration that Eddy's words were God's words, many who studied with Eddy tinkered with what they had learned or grafted it onto some other teacher's theories. William Lyman Johnson, whose father William B. Johnson was one of the fourteen Boston practitioners who remained loyal to Eddy after the Corner schism in 1888, estimated that of the five thousand persons connected with Christian Science in Boston about 1887, fewer than a thousand were followers of Eddy.[9]

Among the other four thousand stood Luther M. Marston, M.D., a spokesperson for many of Boston's generic Christian Scientists, who decried the "unpleasant divisions" that existed among mental healers but also refused to settle for a unity founded on a leader's personal authority. "Our cause will establish itself on a firm foundation, in proportion as personality is left out," he wrote, "and we attain a clearer understanding of the true science of impersonal Mind. Until then, it is the height of folly for any one to claim to be infallible."[10] A. T. Buswell echoed those sentiments before the Boston Convention of Spiritual or Christian Science Healers and Teachers on 25 May 1887. He asserted that "there is no human being on earth whom we should deify for the service of re-enforcing a truth. . . . Our human teachers are simply guides, whoever they may be, and if they claim or allow more, they become as blind leading the blind, and both shall fall into the ditch together."[11]

By the time Marston offered his view of the Corner case to the *Boston Daily Globe*, he had established himself as a major player in the arena of Boston mental healers. A physician who had graduated from a Normal class at Eddy's Massachusetts Metaphysical College in 1884, Marston established the Mental Science and Christian Healing Association in 1886 to organize the efforts of his students "for spiritual endeavor, for uprightness of thinking and living, for peace, harmony and good-will to men."[12] Other institutional efforts followed in quick succession that same year: publication of the inaugural issue of the fifty cents an issue *Mental Healing Monthly* (August), which had 1,800 subscribers by 1888; the opening of the Boston College of Metaphysical Science (December); the founding of the Church of the Divine Unity (December);[13] and the publication of a healing textbook titled *Essentials of Mental Healing: The Theory and Practice*. Marston, who advertised himself as a "Christian Scientist or Mental Physician," regularly published his professional card among other "earnest, honest mental healers" and "qualified teachers of Christian or Mental Science" in the *Mental Healing Monthly*.[14] According to historian Robert Peel's backhanded compliment, Marston's organizations "became the focal (or at least vocal) point for most of the [Eddy] dissidents in Boston—disaffected Christian Scientists, mind-curers, faith healers, magnetic doctors, [and] amalgamators with spiritism and theosophy."[15]

Chicago, the capital of mental healing in the West, presented its own challenges for those struggling to assert Eddy's authority over Christian Science. It was far from Boston's influence, and when Eddy taught her first class there in May 1884, she found Chicago already fertilized by generic Christian Scientists. George B. Charles, B.S., and his wife, Lizzie, who together founded the Illinois Metaphysical College (Therapeutic Metaphysics) in Chicago, illustrated the resultant hybridization of Chicago Christian Science during the 1880s. Edward J. Arens, Eddy's 1881 nemesis, taught "Old Theology" Healing to Charles, who in turn taught Bradford Sherman. Sherman, who began practicing in Chicago in late 1882 or early 1883, took his wife and son to study with Eddy in 1884 and then returned to Chicago to found its orthodox First Church of Christ, Scientist.[16]

By 1886 Marston, anxious about but tolerant of Chicago's metaphysics, which varied from the " 'straight' Metaphysics of the East," identified three schools of mental healing in Chicago by reference to prophetic leadership: Eddy's followers, who are "in earnest and firm in the faith, and Mrs. Eddy is their prophet"; the followers of [Emma Curtis] Hopkins and [Mary H.] Plunkett, who "teach pure metaphysics, but Mrs. Eddy is *not* their prophet"; and the followers of [A. J.] Swarts, who "neither teaches nor talks metaphysics as we understand it, and we imagine he is his own prophet."[17] Even Joseph Adams, founder and editor of the *Chicago Christian Scientist* (1887–90), who arguably championed the unique contributions of Eddy more than any other generic Scientist in Chicago before 1891, wrote that "the Truth taught us at the Massachusetts Metaphysical College, and in Mrs. Eddy's book entitled, 'Science and Health' . . . are still our sources of light, wisdom and power." But, he continued, "all we claim is the privilege of propagating that Truth in our own way, according to our best understanding, ready and grateful to receive suggestions from anyone, but pledging ourselves to nothing which does not come to our own consciousness as the voice of God."[18]

Emma Curtis Hopkins (1849–1925) rapidly assumed leadership responsibilities in the Boston movement after receiving instruction in Eddy's December 1883 Boston class.[19] After a short stint as editor (1884–85) of the *Christian Science Journal*, however, Hopkins broke with Eddy in November 1885 and joined another former Eddy student, Mary H. Plunkett (September class of 1885). Together the two women established the Emma Hopkins College of Christian Science (1886) in Chicago (renamed the Christian Science Theological Seminary in 1887), with Hopkins as teacher and Plunkett as president and promoter, and the Hopkins Metaphysical Association (1886; renamed the Christian Science Association in 1889).[20]

Sometime during late 1887 or early 1888 a rift developed between the two women. Plunkett founded her own school during early winter 1887, published the first issue of *Truth: A Magazine of Christian Science* in November, and accepted her December election as president of the International Christian Science Association in New York. In 1888 she moved her newly named National School of Christian Science to New York and formed an alliance with Marston and Albert Dorman by merging *Truth* (1,700 subscribers) with Marston's *Mental Healing Monthly* (1,800 subscribers) and Dorman's *Messenger of Truth* (1,500 subscribers) into the *International Magazine of Christian Science*, the official organ of the International Association and easily the most widely distributed Christian

Science journal of the time, with a total subscription list of 5,000.[21] Obviously the embarrassment caused by Abby Corner's misfortune was not the only jolt felt by Eddy and her faithful cohort during 1888. Plunkett's leadership proved short-lived, however, as she grew attached to some obscure references to spiritual marriage in *Science and Health* and, while yet married to Mr. Plunkett, entered into "spiritual union" with A. Bently Worthington in 1889. Drummed out of America's mental healing fraternity, the new couple emigrated to New Zealand and founded Students of Truth.

Meanwhile, Hopkins threw her support behind former student Ida A. Nichols's new Chicago publication, *Christian Science* (1888-?), contributed widely to other journals, including the *Christian Metaphysician* (1887–97), and wrote a weekly Bible lesson for the Chicago newspaper *Inter-Ocean* from 1889 until 1898.[22] Through these publications, her seminary (closed in 1895) and association (at least seventeen branches by 1887), and her lectures, Hopkins taught a brand of Christian Science that freely borrowed from a variety of the world's mystical and religious traditions. Eddy denounced such eclecticism, but religious healers who formed New Thought found it invigorating.

Hopkins's lecturing took her from Chicago to both coasts, and she gained a widespread reputation among religious healers as the "teachers' teacher," influencing, among others, Ernest Holmes, founder of Religious Science; Charles and Myrtle Fillmore, who established the Unity School of Christianity; and Malinda E. Cramer and Nona L. Brooks, cofounders of Divine Science. In her later years she retired to New York, where she wrote her major work, *High Mysticism* (1920–22), and conducted private healing lessons.[23]

A second center of Chicago mental healing activity swirled around A. J. Swarts, a former Methodist minister who claimed to have performed mental healings since the 1860s, and his healer wife, Katie L. Swarts, who built their own eclectic school of Mental Science in Chicago by merging the Mind Cure of Warren Felt Evans (1817–89) with Eddy's Christian Science. Evans, himself treated and cured by Quimby in 1863, contributed to mental healing through the publication of such textbooks as *The Mental Cure* (1869) and *The Primitive Mind Cure* (1885) but left the founding of schools and organizations to others like Eddy and Swarts. After Swarts and his wife attended five of the twelve lectures in Eddy's 1884 Chicago course, they launched their own healing work with the October 1884 publication of a new journal titled the *Mind Cure and Science of Life.*[24] After a print battle with Emma Hopkins, then editor of the *Christian Science Journal,* over her charges that Swarts had plagiarized Eddy's writings, the Swartses conducted a successful lecture and teaching tour throughout central Michigan during the summer of 1885 and returned in November to found the Mental Science University in Chicago with George B. Charles, Mary H. Plunkett, and others.[25]

As a result of Swarts's urging, a national gathering of mental healers met during September 1886 in Chicago at the Mental Science National Association. Marston attended from Boston, and many urged him to accept the presidency of the fledgling organization, but he refused. He regretted the "many divisions on the subject of Mental Science" in evidence and noted that "the Metaphysics of the convention were not the 'straight' Metaphysics of the East," but he persisted in his repudiation of autocratic leadership. Marston asserted that "it is the height of folly for any one to claim to be infallible" and hoped that "with a Mental Science National Association in the West, and a National Christian Scien[tist] Association [Eddy's] in the East . . . the cause of Christian Healing will soon be better understood."[26] Swarts, for his part, continued to "condense," "harmonize," and "unify" mental healing by publishing his *Spiritual Healing Formula and Text Book* (1887) for use by any "teachers, students and healers," physicians or otherwise, who accepted the "*scientific statement of Being.*"[27]

Such eclecticism concerned Eddy and her most devoted followers, who tried to make their variety of Christian Science more distinctive and secure by establishing organizations, claiming divine authority, or defining "orthodox" doctrine and standards of behavior. First they fought against the naysayers by telling the story of the historical origins of Christian Science in such a way as to ensure Eddy's intellectual priority and independence. Eddy led in February 1885 by publishing her *Historical Sketch of Metaphysical Healing,* which established her as the direct intellectual descendant of Jesus and the only discoverer and authentic teacher of metaphysical healing in her time. Filled with zeal for her version of truth, she condemned her metaphysical competition as counterfeits and warned the public "that all persons claiming to have been my pupils, who cannot show credentials legally certifying to that effect, are preferring false claims, and are unfit to be employed as metaphysical healers."[28] She followed that blast against her mind healing cousins with a May pamphlet,

Defence of Christian Science, against the "unmerited attack of the Boston clergy" that challenged the authenticity of her Christianity.[29] Second, supporters rose to defend the legitimacy of such claims and to maintain that although her teachings had progressively unfolded, they had not changed in substance and remained internally consistent. As a result, a personality cult began to develop around Eddy.[30] One defender in the *Christian Science Journal* declared: "The founder of Christian Science has no private views to substitute for the teachings of Christianity. Her phraseology and theology are necessary for protection from the above evils [mind cure and malpractice]. Christian Science loses its utility when individuals take it into their own hands, and teach it with their private opinions added thereto."[31] Eddy defended herself against the charge that she had changed her instructions for dealing with "malicious mental malpractice" by claiming that "I endeavor to accommodate my instructions to the present capacity of the learner."[32] A follower proclaimed that "Mrs. Eddy is a prophetess; and she unfolds, like a seer, her mental visions," but "she is not here, or anywhere, to be interpreted as contradicting her own fundamental position and object."[33] The struggle over authentic Christian Science even acquired apocalyptic proportions in the minds of some as Eddy announced that "the great battle of Armageddon is upon us. The powers of evil are leagued together in secret conspiracy against the Lord and against His Christ, as expressed and operative in Christian Science." Others echoed the cry by asserting that the little book the angel commanded Saint John to eat (Rev. 10:9–10) was *Science and Health*, "*sweet* to the spiritual man, but *bitter* to the material man."[34]

Within this maelstrom of debate over the authority of Eddy swirled smaller whirlpools of contention around matters of belief, practice, and organizational structure. What is the true ontology, dualism, or idealism? What is the self? Is Christian Science healing best described as mental healing, mind cure, or metaphysical healing? What are the characteristics of orthodox healing? Do they include the use of affirmations and denials? What are the standards of personal and professional behavior for healers? Is mesmerism a danger or not? What are the proper charges for tuition and treatment? What about specialization? Should Christian Scientists remain as reformers within their Christian churches, or should they come out to form separate organizations? If they should withdraw, what type of organization should they establish, and what should be its mission? Meanwhile, the embarrassingly public deaths in West Medford not only challenged the propriety of practicing metaphysical obstetrics but raised further questions about Eddy's integrity.

The Theory and Practice of Metaphysical Obstetrics

Although childbirth has brought joy to countless women throughout history, for much of that time it represented one of the greatest threats to a woman's life. Not only did many women die from the trauma and infection that often accompanied birth, they lived in fear of the physical pain and debility from labor and delivery and dreaded the often realized prospect that their infants would die during or soon after birth. In the face of these dangers, women had evolved a complex array of cultural habits and social networks to sustain them through labor, delivery, and the often extended period of lying-in. But as historian Judith Walzer Leavitt has so ably demonstrated, since the eighteenth century American women "who could afford them [also] sought out the newest technological and medical advances in obstetrics," with the hope that they "would help them become living mothers of living children." Doubtless many "women sought the promise of improvements, of hope, of 'science,'" but not all found the answer in "science." Some found it in Science.[35]

For years Christian Scientists extolled the advantages of their beliefs for childbirth, and the *Christian Science Journal* regularly published testimonies to their power. Among them was the account of Ella V. Fluno of Lexington, Kentucky, who described in 1887 how her study of *Science and Health* and her confidence in the principles of Christian Science had allowed her to go through pregnancy, birth, and recovery with little pain. She wrote that the principles of Christian Science "so occupied my time that the period of gestation passed with only an occasional symptom of the discords usually attending such periods, and these were destroyed as soon as recognized." She continued many of her normal activities and did not call a birth attendant because her husband had formerly been a physician. When the birth occurred, her husband "said he never witnessed parturition so natural and harmonious. It was almost entirely without a sense of pain. I sat up in bed immediately, and helped to attend the infant." She rose the next day to resume her normal routine. A man who came by the house soon thereafter said that her husband "ought to be prosecuted for allowing his wife to risk her life in such a way," but she and her husband could only sing the praises of Science.[36]

Kate Bigler of Denver successfully delivered her fifth child under the absent treatment of Mrs. Hall and without the presence of a physician. Bigler certified that "the baby was born in a few minutes, with but little pain, and I ate a hearty dinner. . . . I had scarcely any after-pains, was up on the third day, and have done all my housework since the fifth day."[37] Mrs. T. B. of Toronto, Ontario, related that she "had symptoms of the approaching birth and sent for the Scientist who arrived soon, had a very pleasant day, had dinner with my family, and in the afternoon had singing, and reading from Science and Health. During all this time I was not suffering the least pain, and at 7.40 P.M. the child was born, coming without any pain or suffering whatever. This I could not say of two previous births during which I suffered severe pains."[38]

Not all women testified to such routine, pain-free deliveries, but even then Science provided a remedy. Helen Andrews Nixon of Braintree, Massachusetts, recounted that on

the afternoon of July thirty-first occurred a premonitory sign of approaching child-birth. About ten o'clock P.M., after a happy and comfortable afternoon and evening, as I sat listening to the reading of Science and Health, these words seemed spoken to me, "Go lie down, the time has come for the child to be born." I obeyed the "still, small voice" at once. In about fifteen minutes, and without pain, the birth took place. It was what is termed a "dry birth." The "presentation" was the back first, the head being born last, at the third expulsive effort. Without realizing that the proceeding was unusual, I got up at once. . . . I have taken the entire care of the baby night and day. . . . The thoughts centering around such a time have been met before they could be made manifest on the body. In consequence of this, I have experienced no sense of having passed through child-birth.[39]

While these narratives and many more like them attested to the power of Christian Science in some women's lives, they obviously were stylized to advertise the successes of Scientific treatment, especially when regular medicine failed. As a result, however, they also inadvertently tell us which childbirth dangers or difficulties Christian Science women believed their healing system best alleviated. A survey of the childbirth-related testimonies recorded in the *Christian Science Journal* from 1883 to 1910 reveals that mothers most often praised its power to eliminate pain—before, during, or after birth—and they frequently noted a relatively short labor followed by a quick birth and a rapid recovery with only a brief lying-in period. Usually the women simply recounted their experiences and then praised the Truth that rendered such results. However, when a woman on occasion tried to explain her painless delivery, she invariably alluded to the way her confidence in Christian Science teachings about the unreality of anatomy and the nothingness of pain had removed her anxieties and replaced them with a peaceful sense of anticipation. When some students found it difficult to learn that the anatomical and physiological manifestations of pregnancy were illusions, they nonetheless allayed the pain and discomfort associated with childbirth through the soothing "mental" work of the attendant. When, on occasion, women reported that they had summoned physicians or midwives to attend the "mechanics" of the births, they typically added that such assistance had proved unnecessary. According to the testimonies, difficulties rarely presented themselves during pregnancy or childbirth, but when they did, the women claimed successful resolutions to awkward presentations or dangerous complications even when physicians had given up hope.[40]

An unusually detailed account of metaphysical obstetrics appeared in the testimony of Agnes Chester, accused in February 1896 in Kalamazoo, Michigan, of practicing medicine and midwifery in violation of state law. Chester had attended Mrs. Andrew Diehl during her recent childbirth; however, not unlike the Corner case, complications and the advice of an assisting nurse, Isabella Campbell, had led Chester to summon a physician.

Agnes Chester did not attend the Massachusetts Metaphysical College and did not receive any formal instruction in mental midwifery. Although she was aware that Christian Scientists had colleges, she had never attended one and believed that a diploma from one "does not give one an authority over another."[41] Being self-taught, doubtless through her study of Christian Science literature, she "had studied and obtained full knowledge through her own mind," although she conceded that "she had studied anatomy with her brother-in-law, who was a doctor."

She said that she never claimed to practice medicine and had never used medicine or surgical instruments. Referring to herself repeatedly as "a general practitioner in mental science," Chester claimed "on the contrary [that] she was engaged in teaching and healing entirely through the mind." She "testified that Mrs. Diehl had called on her and requested her to treat her on the occasion of the birth of her child." She informed Mrs. Diehl that "her patients could, if they wished, have regular prac[tic]ing physicians in attendance at the same time," but Mrs. Diehl had not wanted a doctor.

When the prosecuting attorney, named Frost, questioned Mrs. Chester regarding her beliefs and treatment, "she explained that it was mental instruction" and repeatedly declared "that she did not treat the disease, but rather the person. She generally went into an adjoining room where she had a better opportunity to concentrate her mind." That way she was not "distracted" by what was happening to the parturient woman. In explaining nurse Campbell's testimony that "Mrs. Chester had made an examination" of Mrs. Diehl, Chester replied that "she made a physical examination of her patients for the reason that the mind and body are one, therefore she had as good a right to examine the latter as the former." She declared that "she did not know of the dangers attended to childbirth, yet was confident in her powers. The body is not material but spiritual substance. God gives birth to the child and is even the immediate cause, using his children as his agencies. There is no such thing as an earthly father. Women are divine ideas; there are no mothers in the earthly sense."

When Frost asked her what were labor pains, she replied that "she did not believe in pain, as she had no knowledge of it. She tried to overcome the false belief regarding the pains of child birth. If the physician understood his business no pains would be known, through a knowledge of the relation between God and the child." She agreed that "there are different presentations at birth" but she believed that one "could foretell them through divine ideas." When Frost asked her how she handled cases of abnormal presentation, her "friends in the audience applauded her answer, 'that she did not have any.' "

As Mrs. Diehl's delivery had proceeded and her condition had grown "worse," Chester had decided based on her "good judgment" that a doctor was needed "for the reason that the medical fraternity would make trouble for her" if she did not. She added that "she knew that physicians were trying to make trouble for her, by her own power of knowing other people's thoughts." When Dr. Pierce, a woman physician, arrived at the Diehl home, she demanded complete charge of the case.[42] Undoubtedly it was Dr. Pierce who reported the case to the medical and legal authorities, who used the case to raise public awareness about the need for more restrictive medical legislation, which finally passed in 1899.[43]

Even before many testimonies to the effectiveness of Christian Science for childbirth began to accumulate, Eddy foresaw the obstetrical potential for her healing system. The wide-ranging advantages included, in her view, not only relief from the sufferings often associated with childbirth, but freedom from the debilitation of bad heredity, a kind of mental eugenics. During her brief stay in Washington, New Hampshire, in early 1882, Eddy distributed a circular that announced her "Parlor Lectures on Practical Metaphysics," which would reveal the way to "improve the moral and physical condition of man to eradicate in children hereditary taints, to enlarge the intellect a hundred per cent., to restore and strengthen memory, to cure consumption, rheumatism, deafness, blindness and every ill the race is heir to."[44] And the announcement highlighted the advantages of her system for childbirth by claiming that she possessed "a certificate from the most celebrated and skillful Obstetrician and Surgeon in Massachusetts, stating our qualification to teach Obstetrics. And what is better, our system prevents the suffering that has attended accouch[e]ment, and with the great auxillary of Mind, obviates the use of medicine."[45]

In preparation for opening her Massachusetts Metaphysical College Eddy had taken a course in obstetrics from Rufus King Noyes, M.D., formerly resident surgeon of the Boston City Hospital and subsequently listed in the 1882 prospectus of the college as "Professor of Surgery and Accouchement." This course, in Eddy's view, certified her with "a requisite knowledge of accouchement," and she advertised herself in several publications as "Professor of Obstetrics, Metaphysics, and Christian Science." Nonetheless, it is unclear whether or not she personally ever used her knowledge of midwifery and anatomy for either practice or instruction.[46] If she did, she must have thought it inadequate, for when a reader of the *Christian Science Journal* asked, " 'Would a true scientist send for an M.D. under any circumstances whatever?' " she answered in the 6 December 1884 issue, "Never except for cases of obstetrics and surgery; only the mechanical operations necessary would be asked of a physician then; restoration to health would be accomplished metaphysically."[47] Notwithstanding this advice, it is clear from *Journal* testimonies that many practitioners did not believe that Eddy's advice *required* the presence of a physician, and most noted that when present his services often proved unnecessary. Either the presentation required no manipulation or, as in the case of Mary H. Philbrick, the practitioner handled it herself. Philbrick reported an obstetrical case in which "surgery was involved, inasmuch as I changed the position of the child from a feet-presentation to the right position. The attending physician said it was wonderful."[48] Similarly, Mary W. Munroe reported a delivery in which "I did the mechanical part for the first time, for lack of others to attend to it. This had always seemed to me something that I never could do. Truly, 'God is Good.'"[49]

Eddy first taught metaphysical obstetrics during one-week sessions in June and December 1887, and immediately thereafter the *Christian Science Journal* printed a flurry of testimonials attesting to the effectiveness of her training. Eugene H. Greene, C.S.B., of Providence, Rhode Island, testified that in his first case after attending Eddy's class in obstetrics "I was not able to carry the belief entirely painless, but nearly so. The patient sat up next day, and walked around the room. People think this a wonderful result."[50] George B. Wickersham of Denver reported that "a lady at Cheyenne, Wyoming Territory, sent me a despatch, asking me to treat her daughter. I did so; and received word from the husband that his wife got along 'just splendidly.' They had no physician at all. She had never done so well before at childbirth. I think this is pretty good for the first case, and an absent treatment."[51] Because Eddy found it a challenge to teach her first class in metaphysical obstetrics, she believed that she had underpriced the tuition. She wrote to Annie V. C. Leavitt, who had missed the June class, that "it is the hardest and the best class I every taught[;] one best suited to the propulsion of the student[.] [I] Have raised my price on it[;] tis now $200 tuition."[52]

Before her 1887 classes Eddy had noted in the March issue of the *Christian Science Journal* that "students are not admitted to the Class in Obstetrics who have not passed through the Primary [class] at this Institution [Massachusetts Metaphysical College]," but she had not explicitly stated that only graduates of such instruction should attend childbirths.[53] Therefore it surprised many of her followers when, under the auspices of the Christian Scientist Association's Committee on Publication, she publicly denounced Abby Corner in the Boston newspapers as a quack and impostor. According to Eddy, Mrs. Corner never entered the obstetric class at the Massachusetts Metaphysical College. She was not fitted at this institute for an accouche[u]r; had attended but one term, and four terms, including three years of successful practice by the student, are required to complete the college course. No student graduates under four years. Mrs. Eddy, president of this college, requires her students to use the utmost precaution in practice, and to be thoroughly qualified for their work. Hence the rapid growth of this system of mind-healing, its safety and success. The West Medford case, so far as is known, is the first instance of death at child-birth in the practice of Christian Science. This fact is of vital importance when compared with the daily statistics of death on such occasions, caused by the use of drugs and instruments.[54]

Although many of the details remain hidden, the Corner experience appeared to bring to a boil issues that had been simmering for some time. A close follower of Eddy's, probably J. M. C. Murphy (who attended her 6 June 1887 class in obstetrics) or W. H. Bertram, had consulted her about the advisability of taking an obstetrics course at a "material" medical school. She had agreed, even going so far as to offer him financial assistance, "provided he received these lessons of a certain regular-school physician, whose instructions included about twelve lessons, three weeks' time, and the surgical part of midwifery." After he had completed the course, he asked Eddy if it might not be a good idea to enter medical school, to which she later reported she had "objected on the ground that it was inconsistent with Christian Science" but that, "notwithstanding my objection, he should do as he deemed best, for I claim no jurisdiction over any students."[55] In the March 1888 issue of the *Christian Science Journal* Eddy tried to differentiate her view of the appropriate relationship between material medicine and metaphysical healing from the views of other mental healers:

Homoeopathy is the last link in material medicine. The next step is medicine in Mind. One of the foremost virtues of homoeopathy is the exclusion of compounds from its pharmacy. I wish the students of Christian Science (and many who are not students understand enough of this matter to heed the advice) to keep out of their heads the notion that compounded metaphysics (so-called) is, or can be, Christian Science. They should take our magazine, work for it, and read it. They should eschew all magazines and books which are less than the best.[56]

Despite such advice against mixing mental and material medicine, several Christian Scientists entered a three-year program to earn their M.D. degrees.[57]

Sarah Crosse, Abby Corner's relative and a member of the Committee on Publication, and her husband took the lead in defending Corner against Eddy's public attack. In their view Corner, like many practitioners before her, had simply assisted a childbirth and had summoned a physician when the situation got complicated; but when things went terribly wrong, Eddy had turned on Corner in an attempt to protect her reputation and that of her college. Joining the Crosses and others, Murphy raised $170 in commitments from association members for Corner's legal defense. A struggle for control of the association ensued, with Calvin Frye, a longtime servant of Eddy's, warning that "there is a huge conspiracy in our midst, and appearances implicate Mrs & Mr Crosse; H. P. Bailey; J M C Murphy; W. H. Bertram; and I fear even Br. Troup & Harris, I tell you this that you may know who is trying to black our leader and who are her and our enemies."[58] By the end of June one-third of Eddy's Boston followers, disillusioned and disaffected, had left her church, thereby decimating the leadership.

Although undeterred by recent events, Eddy proceeded more cautiously in October when she offered another obstetrics course. She co-taught the course with Ebenezer J. Foster, a homeopathic physician who had converted to Christian Science and who later became her adopted son. Foster instructed the students in the anatomy and physiology of childbirth during the first five sessions, and Eddy concluded the course with four lessons on its metaphysics.[59] We do not know exactly what Foster and Eddy taught, but some hints appear in the surviving lecture notes of Alfred E. Baker, M.D., C.S.D. (d. 1924), whom Eddy called to teach several obstetrics classes from 1899 to 1901 under the auspices of the Board of Education.[60]

Thirty-three students registered for Baker's 1900 class, which began with the general reminder that "error disappears as we recognize that the creator, the one Mind, governs all and takes care of all."[61] He then proceeded to give the standard Christian Science denial of the material meanings of words such as anatomy, physiology, and chemistry, to remind his listeners of their nothingness, and to encourage them to review Eddy's "references on anatomy, physiology, embryology" and "bones," which affirm that "man is not structural."[62] Given these premises, it followed that the male and female reproductive organs and the processes of intercourse and conception also had their spiritual reality. Because there are no distinctive male or female organs and because nerves do not exist, there is "no sexual desire or genital sense. . . . The capacity to reflect the eternal is intercourse. That is Love." The physical union of marriage, in which, so often, the man has "pleasure and the woman pain," is an error of belief that causes "much dissension" and should be remedied by an affirmation of Eddy's recognition that "'unity of Principle and idea is the only marriage.'"[63] Sperm and ovum do not physically conjoin in conception, for sperm "is Truth, Mind, and the expression of Mind" and the "seed of God is the angels," and women should lay aside the "superstition" that "Menstrual function" accompanies "egg formation."[64] Apparently Eddy herself believed that she had "overcome" menstruation in 1866, for she claimed in 1903 that "the monthly period left me the moment I came to C. S. and I have never seen anything of the claim since."[65]

The woman's pelvis is a "shadow picture" that will be outgrown and cannot be deformed or become rigid and there are no material organs within it. Therefore, a uterus is simply a "place where anything is generated or produced" and since "Mind is the only producer," each of us must choose what is produced.[66] True birth is not the expulsion of a child from a woman's pelvis, but "revelation," "understanding," and the act of "receiving Life, Truth, Love, [and] action, all the time." It is the manifestation of God, the one Father-Mother and "the woman out-living the finite mortal definition of herself." Eventually, a woman who grows in Truth will recognize that "there is one Mother, even God, and will lose all desire for creation apart from God, Love, and then comes the millennium." In sum:

> Birth is painless, as the flowers and bees. We put the stings into bees. Birth is lustless; [it] is a manifestation of Life, Truth, [and] Love. When man sees birth right, he will see no pain, no sensation or pleasure in matter, but will see God, the oneness of Mind. . . . Hold this thought in treatment, it will heal. There is no material man, no material woman, no sex. Mortal men are material falsities. [There are] no male or female

generative organs, no conception—painful or otherwise—no embryonic development.[67]

Practitioners should make mental preparations before attending a childbirth and remind themselves that "the way to heal is to know there is no one to heal" and that "we only heal out of our own perfection by reflecting the one Mind and living the life of Christ." The healer should carefully study pertinent passages in *Science and Health*, including the "first four pages" of the chapter "Christian Science Healing," and reflect on Psalm 91, which is a meditation on God as the protector and deliverer of those who find refuge in him.[68] Finally, throughout their relationships with a patient, practitioners "must reflect compassion to deliver the patient." If the healer is not "rid of self," then she has "more to be delivered from than to deliver the woman." Remember that "if you don't have a living interest in patients, [you] never can heal them."[69] Often this spirit of goodwill is especially necessary when a physician is also in attendance at the birth. Then the healer should let the physician know that a Scientist is engaged in her silent work and "abide in the consciousness of action. Action is because God is. God takes care of it."[70]

Using Baker's notes, we can examine the activities of a practitioner in normal cases—from pregnancy and the creation of the proper birthing environment through labor, delivery, and lying-in. Although at times Baker gave specific suggestions for care, he cautioned his students against applying his or Eddy's instructions to patients like formulas or magic potions: "Formulas are narcotics. Mother [Eddy] said [we] might as well give drugs as formulas." Many generic Christian Scientists, including Swarts and the students of Hopkins, published manuals that summarized healing step-by-step. According to Baker, although at times *Science and Health* may appear to be formulaic, it "is a manifestation of light, not a formula. . . . You can no more formulate this book than you can put God into your pocket, or in matter."[71]

Baker highlighted general helps for pregnant women that would allay their fears and prepare them to understand the true nature of childbirth. They should read and meditate upon the spiritual meanings of Psalms 91 and 23, Jesus' Sermon on the Mount, and *Science and Health*. Although the "scientific statement of being" must be used with the right thought, it was also excellent. Finally, references in Eddy's writings to the following words could be helpful: bride, bridegroom, Adam, ark, children, children of Israel, creator, Eve, father, mother, matter, mercy, mortal mind, Mind, resurrection, salvation, temple, uncleanliness, and wine. Morning sickness, the bane of many expectant mothers, was to be met with the recognition that there are no nerves and hence "no reflex action to cause nausea."[72]

What one might call the birthing environment also needed the attention of the healer. Concerned that practitioners not, with a fanatical application of metaphysics, ignore ordinary habits of cleanliness, Baker urged his students to remember that "it is well to have [a] bright sunny room and fresh air." There are "no microbes, nor organic life, or living organisms." However, "these things must be handled—like sweeping a room or baking a cake—with hands when necessary. Keep things clean."[73]

But more than the "material" surroundings needed to be clean; the world of thought needed to be mentally decontaminated. Sometimes expectant parents might think that they do not want children, and they might dwell on such thoughts in an effort to abort a fetus. The practitioner should handle abortion thoughts by reminding the parents that they must accept the consequences of their union and by explaining that such thinking indicates "a thought of murder back of stopping [the] development of [the] embryo." If the woman's husband appeared not to want his wife to have a Christian Scientist attending the birth, then the practitioner must meet in herself the "unwilling husband thought" by "overcoming self-justification or human reason, superstition and idolatry—one of the worst sins to be overcome" and remind herself that "there was no unwilling husband to meet there." If the practitioner failed to meet this claim, the results would be a "wil[l]ful child" and a "belief of obstruction" during delivery. Neither the parturient nor the practitioner should worry about what the mother-in-law thought. "Mother-in-law thought needs to be handled" by remembering that there is "no mother-in-law, just mother-in-love." A practitioner should not give any thought to a family's state of poverty or wealth. In fact, if she were going to allow conditions of poverty or squalor to impress her, she "might as well stay away," for surely "if money is anything, God will melt it." As labor approached, rather than fear that animal magnetism is waiting to consume the child after birth, the practitioner should meet the old red dragon, error, and handle the serpent as did Moses of old.[74]

The Christian Scientist should recognize that the onset of labor signals "no change of system," no pain nor exhaustion for the woman, and the practitioner should not physically examine her, for " 'matter disappears under the microscope of Spirit.' " Eddy asserted that the pain of childbirth "consists in nature trying to expel what has become too large," but since there is no pain in conception there can be none in birth. Work to "disperse the belief that a doctor can carry a belief of pain" and "realize that woman is already delivered." There should be no fear of bleeding or of a retained afterbirth, for "'perfect Love casteth out fear,' and placenta and everything." The practitioner must guard her thoughts on seeing the newborn lest false claims about its sex, its beauty, or any prenatal influences negatively affect it. Do not worry if the baby fails to nurse: "Compassion brings out [the] milk supply," and anyway there can be "no loss to [a] child because of [a] lack of nutrition." Finally, although Scientists knew they could tell the new mother to "get right up" after the birth of her baby, they should not force the issue, especially in front of nonbelieving attendants.[75]

Baker included specific instructions for handling the false claims of births, what one might call physical complications, including dry birth; abortion (stillbirth?); pressure, overaction, or clot of the mother's bowels; and rupture of the peritoneum. According to his instructions, an effective treatment for reversal obstruction, the "leading error to be handled," was Eddy's hymn 298 in the *Christian Science Hymnal*. The fourth stanza of the hymn reads:

> *Strongest deliverer,*
> *friend of the friendless*
> *Life of all being divine:*
> *Thou the Christ, and not the creed;*
> *Thou the Truth in thought and deed;*
> *Thou the water, the bread, and the wine.*[76]

In Baker's view there was no such thing as a strangulated umbilical cord; therefore a Christian Science nurse in attendance should not touch the umbilical cord. He also addressed the false claims of childbed fever or puerperal mania, which in his view would occur only if the practitioner neglected her patient and failed to love, lead, and cherish.[77] Finally, he asserted the nonexistence of syphilis, locomotor ataxia, scrofula, and lead poisoning.[78]

Just what difficulties or successes metaphysical obstetricians discovered when they incorporated Baker's detailed instructions into their practices is unknown, and no one taught the course again. The April 1902 issue of the *Christian Science Journal* announced an amendment to the bylaws of the *Church Manual:* "Obstetrics is not Science, and will not be taught."[79] With that terse sentence Eddy discontinued specialized instruction in obstetrics, which included the study of anatomy and manipulation, and thereby eliminated an awkward exception to material-free Christian Science instruction.

The Routinization of Eddy's Authority

Driven from Boston by the turmoil and defection surrounding the Corner case and the chaotic clamorings of mental healers of all stripes, in May 1889 Eddy retreated to Concord, New Hampshire, to receive God's instructions for the future of Christian Science. By then she had become in the eyes of some a virtual Urim and Thummim for resolving disputes or offering guidance, and the *Christian Science Journal* begged its readers to allow her to commune with God in peace and to "let us keep from her and settle among ourselves, or, with God for ourselves, the small concerns for which we have looked to her."[80] In the view of others, she had through selfish ambition assumed for herself a divine calling that no mortal should claim. When she reemerged from her seclusion, orthodox Christian Scientists used Eddy's prophetic pronouncements to create a locus around which believers and practitioners could formulate policies regarding the community's relation to American culture. Beginning with the Massachusetts Metaphysical College, which had become a lightning rod for much of the discord surrounding the Corner case, she dismantled the organizations that had grown up around her but that had increasingly expanded or wandered beyond her control.[81] She dissolved the Christian Scientist Association, disbanded the Church of Christ, Scientist, and temporarily adjourned the National Christian Scientist Association. Recognizing the limitations of her followers, she stopped short, however, of completely spiritualizing the movement. She cautioned in December 1889:

> I do not require Christian Scientists to stop teaching, to dissolve their organizations, or to desist from organizing churches and associations.
> The first and only College of Christian Science Mind-healing, after

accomplishing the greatest work of the ages and at the pinnacle of prosperity, is closed. . . .

When students have fulfilled all the good ends of organization and are convinced that by leaving the material forms thereof a higher spiritual unity is won, then is the time to follow the example of the *Alma Mater.* Material organization is requisite in the beginning, but when it has done its work, the purely Christly method of teaching and preaching must be adopted. On the same principle you continue the mental argument in the practice of Christian healing until you can cure without it instantaneously and through Spirit alone.[82]

For years Eddy had claimed to heal instantaneously, and now she seemed to be saying that she had also matured beyond needing colleges for teaching; she could now teach and preach like Christ. She may have been right, but it is also clear that by dismantling the old organizational structures of her movement, through which she had so visibly but unevenly struggled against false Science, she effectively purged her system of most "pseudo" Scientists. When she invented "the purely Christly method of teaching and preaching" three years later, she permitted only "true" Scientists to enter.

The last fifteen years of Eddy's life (1895–1910) coincided with a period of rapid growth for Christian Science. Much of this growth undoubtedly resulted from the reinvigorated organization Eddy established, but it also followed efforts by the movement's leaders to foster a more "mainstream" public image. Internal challenges to Eddy's prophetic authority still occasionally surfaced within the movement, and she expended great energy to consolidate and extend her authority. By and large, however, her followers held her in great esteem and turned their attention to efforts to prove themselves to the outside world through evangelism and professionalization.

In the fall of 1892, Eddy appointed four lieutenants to execute her plans for the institutional reorganization of Christian Science. As members of the Christian Science Board of Directors, they began construction in Boston on a new central church headquarters, the Mother Church, to which all truly orthodox Scientists would belong. Throughout the 1890s Eddy solidified her doctrinal authority over members of the Mother Church, exerted discipline through advice or threats of excommunication, and sought to ensure the orthodoxy of Christian Science teachers. Asserting that "the Bible was my only text-book" for the preparation of *Science and Health* and denouncing critics who charged otherwise, she reaffirmed her claims to divine insight and literary independence.[83] In 1894 she ordained the Bible and *Science and Health* as church pastor. Charismatic human pastors posed the potential danger of leading congregations away from Eddy and orthodox Science through their sermons. But an "impersonal pastor," from which assigned passages would simply be read and not commented on at church services, would encourage doctrinal uniformity and decrease the chances that any individuals could seriously challenge the privileged relationship among Eddy, her writings, and their orthodox interpretation.[84]

To keep her instructions updated and in demand, Eddy regularly revised *Science and Health,* and in 1895 she published the first edition of a *Church Manual* that "became the ultimate authority for all action by the church" and codified her numerous bylaws and instructions.[85] Given the essential connection for Scientists between right thinking and good health, one can sense the importance of supplying orthodox literature for the protection of its members and the satisfaction of an inquiring public. Through the Christian Science Publishing Society, established in 1898, the church met this need by assuming responsibility for the publication of the *Christian Science Journal* and *Christian Science Sentinel* (official organs of the Mother Church), the international *Christian Science Monitor* (1908), and numerous other printed materials that bore witness to the message of Christian Science. Eddy completed her hegemony over orthodox Christian Science in 1898 by replacing her students' regional institutes with a centralized Board of Education to direct the instruction of new teachers and to monitor their orthodoxy.

Reasonably confident that she had secured the internal structure of her church, Eddy turned to generating goodwill and converts for Christian Science. To this end she orchestrated efforts, beyond the everyday witness of believers and practitioners, to induce the public to look favorably on her movement. Lecturers held well-advertised "evangelistic" meetings across the country for nonbelievers, and one-person Committees on Publication defended Christian Science beliefs or behavior and sought to defuse bad publicity by correcting public "misunderstandings" in the press. Both committees and lecturers received close supervision from regulative boards of the Mother Church.

Unwilling to conform to the dictates of Eddy, many of the generic Christian Scientists became part of the loosely constituted New Thought movement. Participants in this movement variously practiced mind cure, mental cure, or metaphysical healing, but they shared a conviction that the mind can solve all human problems and a commitment to a highly idealist philosophy drawn from such writings as those of the eighteenth-century Swedish mystic Emanuel Swedenborg, Asian and Judeo-Christian teachers, and nineteenth-century spiritualists.[86] After several attempts to establish a unified organization, generic Christian Scientists and New Thoughters drew on their obvious affinities and formed the International New Thought Alliance in 1914. Today the Unity School of Christianity, Religious Science, and Divine Science organizations are the nearest relatives to the New Thought and generic Christian Science movements. Members of these organizations continue to search for the harmony of being that reveals itself through healthier and more successful lives, but unlike most orthodox members of Eddy's Church of Christ, Scientist, they will consult practitioners of scientific medicine to augment mind healing. The Unity School of Christianity widely influences mind healers today through its healing journal *Unity* (1891), with an annual circulation of over 200,000 copies, and Religious Science and Divine Science together claim nearly three hundred churches and over six hundred practitioners in the United States.

PART 2

Christian Science Healers
and the World

5

Physicians Debate Christian Science

Whatever may have been its origin, Christian science, or the mind-cure, is now the most popular craze in some of our large cities; and, like spiritualism, like the skating-rink, and like the airs of "Pinafore," will probably sweep over the country until it has penetrated to the remotest hamlet, and may found a school of medicine as popular as that of Hahnemann.
—Mary J. Finley, M.D., 1887

The whole movement of Christian Science is a natural protest against the materialistic trend of modern medicine.
—Frank B. Wynn, M.D., 1921

From their earliest years Christian Scientists, struggling against the principles and practices of medicine, paid close attention to the American medical community; but organized medicine hardly noticed Christian Science until the late 1880s. Even then only a few physicians, like Ohio's Mary J. Finley, feared that Scientists might develop into "a school of medicine as popular as that of [Samuel Christian] Hahnemann," the founder of homeopathy. Many more viewed the movement as nothing more than an especially odd variety of mind cure, faith cure, or religious healing; and only much later did a handful, like Indiana's Frank B. Wynn, view it as "a natural protest against the materialistic trend of modern medicine." However, as the membership growth of Christian Science accelerated after 1890, the medical community commenced a vigorous discussion in its medical journals about the new medicoreligious movement. Despite their initial assessments, physicians quickly recognized that they needed to understand more about these new healers who might threaten their income as well as the public's well-being (see figs. 1 and 2).

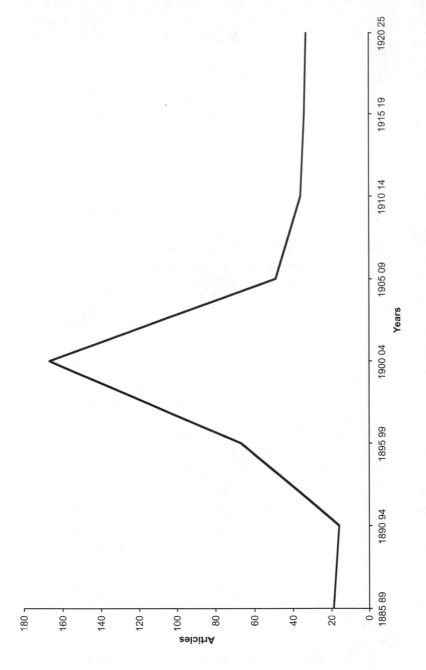

Figure 1. Medical Journal Articles about Christian Science, 1885–1925

By the late 1890s the medical press was assuming that Christian Scientists numbered one million in the United States, and since in theory every adult Scientist could be considered a practitioner, such an army of competitors could have presented a real threat to regular medicine.[1] The reality, of course, was quite different. True, the decades around the turn of the century witnessed a dramatic jump in Christian Science membership, and the number of full-time practitioners grew to rival that of the eclectics and homeopaths in the nineteenth century and the osteopaths and chiropractors in the twentieth, but Scientists still accounted for only about 5 percent of the estimated one million.[2] Whether physicians intentionally inflated the figures to exaggerate the Christian Science threat or were simply ill informed is not clear; in any case, many of them perceived Scientists as a serious menace to their efforts to persuade Americans that they were the best arbiters regarding matters of health.

An examination of American medical periodicals reveals that the debate focused on three questions: What is Christian Science? Does Christian Science heal people? How can physicians absorb into their own practices the good of Christian Science while eradicating or curtailing the rest? In each case physicians structured their answers to trivialize the meaning and influence of Christian Science while enhancing their own self-esteem and strengthening their stature in the public eye.

What Is It? Definitions of Christian Science

As the medical community in Progressive Era America discussed the theories and practices of Christian Scientists, it worked to control the terms of the debate so that Scientists would appear marginal and foolish. Physicians defined Christian Science as a blatant example of the many past and present popular systems of religious and mental healing, quackery, and deceit, each posing a menace to the health and morals of an unwitting public. In addition, they often depicted Christian Scientists as a threat to their own status and income. By adopting such attitudes toward Christian Science, physicians unwittingly displayed both disdain for the public's judgment and anxiety about their own authority.

For many physicians, Christian Science represented the most popular and egregious example of an upsurge of contemporary interest in mind cure and religious healing that scientific medicine struggled with after about 1885. As physicians searched for an explanation of this grow-

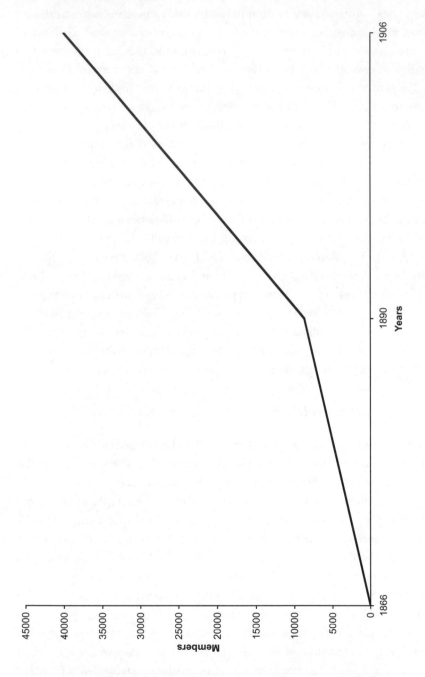

Figure 2. Christian Science membership, 1866–1906

ing popularity, some offered the obvious answer that Christian Science thrived because it worked, especially for those whose illness contained a mental component. William A. White, superintendent of the Government Hospital for the Insane in Washington, D.C., believed that the medicoreligious sects grew "because of a need which a large number of people have, and they are, so to speak, the answer to that need. That need, of course, expressed simply is the need for better health."[3] Although such an explanation must have held some attraction because of its simplicity, it had the liability of implying that the public knew best. If doctors agreed, they could become simple empirics who chose treatment plans based on popularity polls.

Other physicians, preferring to blame the popularity of Christian Science on deceitful practitioners, viewed it as the latest version of quackery, the perennial bugbear of all reputable healers, and derisively compared it to Perkins's tractors of an earlier century.[4] Similarly, some blamed it on a resurgence of ancient magic, medieval witchcraft, and religious superstition under the direction of modern-day con artists. Thus understood, the popularity of Christian Science served as a cautionary tale for physicians about the continued credulity of the masses even in the face of the advances of modern medicine. In this vein St. Louis physician Alexander B. Shaw, speaking in 1889, attributed Christian Science "to a co-working of imagination, credulity and superstition," sentiments echoed a decade later by Chicago physician James G. Kiernan when he wrote of "confidence operators and instabilities."[5]

Those more charitable to religious healing in general saw Christian Science as a distorted recurrence of the healing practices of premodern cultures or of a kind with the efforts of Dowieites and other biblical literalists to resurrect the healing practices of New Testament times. One of the best examples of some commentators' tendency to view Christian Science as a modern example of a mythopoeic worldview appeared in the *New York Medical Journal*. Written by Henry K. Craig, curator of the Pathological Museum of the George Washington Medical School in Washington, D.C., the article traced the worldwide history of popular mythology, with particular attention to the role of the dreaming state in popular myth. Craig concluded that in the Christian Science attempt to transform

the natural man into the supernatural by a mental process, the nineteenth century witnessed the reproduction of a philosophy not less than three thousand years old. And the essential difference between the dream of the Hindu and the dream of the Eddyite?—the Hindu accomplished his object by self torture, the Eddyite attains his end by a method which, if the mental anguish of the generality of readers is placed in the balance, surpasses the entire catalogue of Hindu cruelties, namely the reading of Mrs. Eddy's writings.[6]

But in a more typical historical analysis that appeared in 1904, Edward Wallace Lee, a physician from New York City, blamed the public. "There were," he admitted, "even in the most ancient times, educated men; but the masses were ignorant. The same conditions obtain to-day: the masses are ignorant, and are more willing to accept some incomprehensible, pseudophilosophy than plain, simple truth."[7]

Avoiding such generalized critiques, Mary J. Finley as early as 1887 called on the members of the North Central Ohio Medical Society to examine the *specific* theories and practices of Christian Scientists.[8] Until the turn of the century, however, most members of the medical community, like the editors of the *Journal of the American Medical Association* (*JAMA*), could not have cared less whether a religious healer was one of "the self-styled 'Christian Scientists'" or a follower "of the kindred though rival delusion known as Dowieism," because "each is alike damnable in every sanitary point of view."[9] In this view the beliefs of the groups mattered little compared with the dangerous consequences of their practices. Those physicians who tried to distinguish among the various healing movements and to analyze the specific history, beliefs, and practices of Christian Scientists varied in their commitment to treat the movement fairly. But they at least recognized it as something worthy of attention in its own right, if for no other reason than that it posed a distinct threat.

Those physicians who did try to understand the movement on its own terms generally began with an examination of the life, character, and personality of Mary Baker Eddy and then turned to an analysis of the distinctive name "Christian Science." They usually "discovered" an unscrupulous and mentally unbalanced woman who had founded a well-financed and well-oiled organization for the sole purpose of acquiring wealth and power.[10] In short, they found, as *JAMA* regularly put it, a "profitess" in the guise of a prophetess.[11] And after Georgine Milmine's exposé of Eddy appeared in *McClure's Magazine* (1907–8), they also "found" a psychologically disturbed woman. Using the diagnostic terms of the day, *JAMA* judged her a "hysteric" and "psycho-neurotic" who exhibited "classical hysterical paraplegia." Ralph Wallace Reed, alienist and neurologist to the Cincinnati Polyclinic and Post-Graduate Medical School, claimed that she exhibited "major hysteria," "psychosis," and "involutional paranoia."[12] For most of these commentators the real tragedy was that Eddy's followers were people in genuine emotional and physical need who had been duped for selfish ends. Thus told, the history of Christian Science became a sort of morality play about the extremes to which false religious leaders will go to enslave their followers for commercial gain.

Although not as titillating as armchair psychology, the study of the name "Christian Science" proved more problematic. With that name, Christian Scientists announced their belief that they offered a new scientific Christianity for a desperate world. This made it easy for even the public to recognize the hybrid nature of the movement as both religious and medical in its orientation, something the medical community noted with great consternation. Edmund Andrews, professor of clinical surgery at Northwestern University Medical School, observed with some concern in 1899 that Christian Science "intrudes itself upon us both as a religion and healing art inextricably mixed together. Moreover, it uses its peculiar theology as a weapon to attack the science of medicine, and with it all the other physical sciences."[13]

Whereas postbellum American physicians had shifted the base for their professional identity from a matter of character to one of expert knowledge about medical science, their view of themselves as community leaders, if not always family practitioners, led them still to speak out on religious and moral issues.[14] In fact some, including St. Louis physician J. M. Bradley, believed that because a physician treats the whole person and "recognizes that as the states of the body may affect the diseases of the soul, so also the conditions of the mind influence the states and diseases of the body," then he is better suited than the philosopher or clergyman to speak on such matters.[15] It was not strange, then, that physicians asked not only whether Christian Science was *really* scientific, but also whether it was *really* Christian. In both cases they usually answered no.

In the view of many physicians, Christian Scientists neither believed nor practiced true Christianity. In 1890 the secretary of the Iowa state board of health, J. F. Kennedy, concluded that Christian Science is "Satanic in origin, fiendish in practice, and murderous in results." And many agreed with the view if not the florid prose of Cincinnati's James B. Taylor: "It is pantheistic in the clearest sense. It is optimism run mad. It is the most crazy stage and application of idealism. 'Pantheo-optimo-idealistic Practice' should be its name. It is not Christian."[16] The medical community might have enlisted the expertise of the clergy for a religious critique of Christian Science, but only rarely did an article about Christian Science by a clergyman appear in a medical journal between 1880 and 1920.[17] This does not mean the journals completely ignored the views of the clerics; occasionally editors reported their negative opinions, and at least once *JAMA* published a list of recommended readings on Christian Science that included several books by ministers.[18] However, these were the exceptions. This reticence may have occurred in part because physicians did not know if the clergy were their allies or their enemies in health matters. Edward Wallace Lee wrote in 1904 that "some ministers of the gospel are too prone to give their endorsement to quack nostrums whose ingredients and effect they know positively nothing about," and Philip Mills Jones, activist editor of the *California State Journal of Medicine*, concurred in 1909 when, in criticism of the growing Emmanuel Movement he wrote that "there has always been a subtle relation between medical vagaries and the ministry; it is said on good authority that a very large percentage of the notorious quacks in the United States are broken down or unfrocked clergymen. Facts are facts and the elemental type of the average human mind is one of them that can not be ignored."[19]

Whereas Eddy herself and the vaguely iconoclastic name "Christian Science" raised general concerns among physician critics, the Christian Science doctrine that denied the existence of matter and disease provoked outright contempt. Not only did such a belief take away a physician's raison d'être, it flew in the face of empirical science and common sense.[20] What, wondered the editors of *JAMA*, would a Christian Scientist do if a carload of bricks fell on a child? Undoubtedly he would remove them, but why would he not remove the toxins that have similarly befallen a child stricken with diphtheria?[21] If such toxins are an illusion, reasoned an unidentified physician, then Christian Scientists should "permit someone to inject them with the germs of diphtheria, or tetanus, or gonorrhea, or to take a 5-grain dose of morphine or strychnine, or atropine."[22] Although such a test might have been viewed as casting pearls before swine,

William D. McCracken, a Christian Science lecturer speaking at the Metropolitan Opera House in St. Paul, Minnesota, accepted the challenge in principle when he publicly offered to drink germ-laden water with the confident declaration that no ill effects would follow. "I scoff at the ideas of Dr. Pasteur, and I do not believe that such fanatics believe what they teach," said McCracken, for "there are no germs and water is the same, if you have faith, whether it is boiled, filtered or taken raw."[23]

Physicians themselves had engaged in such debates about the etiology of disease countless times in the past, and McCracken's challenge could have been made by any number of contemporary antigerm theorists. However, Christian Scientists denied not just "invisible" germs but the whole biomechanical model of disease that underlay claims to their existence. If true, Christian Science metaphysics meant that physicians would be unnecessary in the near future; they would be unemployed on earth as well as in heaven.

For many physicians, such metaphysical meanderings seemed little more than religious mumbo jumbo. But for those who followed the reasoning of Christian Scientists, it seemed patent nonsense for Scientists to deny the existence of disease and then turn right around and claim they had cured it. "It must be very interesting," wrote Northwestern's Edmund Andrews, "to see a non-existent bone, having only the substantiality of thought, which had been dislocated from a socket which never existed, going back without being touched into that imaginary socket with a thump audible to all the bystanders."[24]

To add insult to injury, Christian Scientists charged patients for treatments while claiming to practice the religious healing of Jesus. *JAMA* sarcastically hypothesized that "the unreality of a germ as distinguished from the materiality, let us say, of the two dollar bill which a Christian Science practitioner exacts for an 'absent treatment' is based, presumably, on the fact that one can be seen only through a microscope while the other is visible to the naked eye."[25] On the one hand, physicians believed that practitioners were selfish to charge for what amounted to no more than a prayer for the sick and criticized them for not more actively ministering among the poor.[26] On the other hand, they begrudged a practitioner's low cost of doing business by "absent treatments," which required no "outlay of time or money for study, equipment, office rent, etc."[27] They also believed that Scientists attracted many patients because "the usual fee is one dollar a 'treatment'" and there is a "total absence of all expense for medicine."[28] Christian Science seemed to be primarily a system of medical treatment masquerading under the veil of a misguided religion.

Only slim evidence survives that the intensity of the physicians' protests against "payment for miracles" resulted from their fear that Christian Science practitioners were costing them patients and income. W. S. Turner, an eclectic physician from Ohio, voiced his concerns at the 1909 state Eclectic Medical Association meetings that "osteopathy, Christian Science and other similar creeds" had "begun to draw heavily on the clientele of the regular profession."[29] And in 1915 Samuel A. Tannenbaum complained to his colleagues at the New York City Medical Alliance that "the patients we lose through the activities of the Eddyites are the fairly prosperous middle class—lawyers, actors, school-teachers, writers, amateur art enthusiasts, and the nouveaux riches. . . . Eddyism does not flourish among the poor."[30] There undoubtedly existed local or regional evidence for such concerns, and Christian Scientists may have made significantly greater inroads into the practices of medical sectarians, such as eclectics, than of regular physicians. But on a national scale, the grounds for such fears were probably as unreal as the fantastic physician estimates of Christian Science membership. John Ferguson, a Toronto physician, had a more typical appraisal of the effect of Christian Science on medical practice when he argued that "no false system of treatment can in the least degree interfere with the amount of cases that must come the way of the regular medical practitioners. Indeed, as a matter of fact, these false systems can have no other result than that of increasing the sum total of disease and suffering."[31] J. E. Engstad, M.D., from Grand Forks, North Dakota, also discounted the belief that "Christian Science has directly curtailed the income of the physician."[32]

Laying aside all the confusing doctrines and questionable practices of Christian Scientists, many physicians concluded that the refusal of Christian Scientists to practice healing without charging for their services proved they were not simply religious healers. Scientists, working under the false colors of religious healing, had gained a modicum of respect, but on close analysis one discovered, in the words of Samuel A. Tannenbaum, that "Eddyism is the practice of medicine under the guise of a religion."[33] Andrews summed up the "scheme" by noting that Christian Science had "come to the front at a favorable time. Eclecticism and homeopathy have lost their sharp separateness and are seeking to merge themselves into general medicine. This leaves a vacancy, and Christian Science promptly steps in to fill it."[34]

Did Patients Get Well? A Matter of Evidence

Christian Scientists claimed that American mortality rates had decreased as a direct result of the spread of their practices and asserted that believers enjoyed much healthier and longer lives than nonbelievers. For example, Scientist Richard P. Verrall maintained that the New York City Central Park West church had "a membership of over thirteen hundred, among whom the mortality for 1903 was a quarter of one per cent"; he neglected, however, to state the city's overall rate or to acknowledge that factors other than the practice of Christian Science may have influenced mortality.[35] And Peter V. Ross, the Christian Science Committee on Publication for northern California, claimed that during the great influenza pandemic of 1918–19 Christian Science had "quietly healed thousands who turned to it for help, while modern medicine stood by helpless, ignorant alike of the cause, cure and prevention of the malady."[36] Such claims may have served in part to blunt the influence of critics who claimed that "Christian Science statistics do not exist in any valuable and detailed form, because the healers do not examine and diagnose diseases, but simply deny their existence."[37] But the assertions did little to assuage the doubts of the medical community.

As early as 1888 James B. Taylor had observed that "while there are occasional striking phenomena, the reported triumphs of the school are . . . usually fallacious as to the fact of cure (often erroneously claimed), the nature of the malady (of which the 'scientist' is no judge), or the cause of relief (every good office of nature being in their eyes the product of their science)."[38] According to doctors, Scientists primarily used unexamined testimonies to establish efficacy and failed to provide trustworthy statistical evidence for their cure rates.[39] In 1909 Chicago physician Alfred D. Kohn noted that the Christian Science textbooks and periodicals were "full of testimonials of wonderful cures of disease, which sound much like patent medicine advertisements in the newspapers," but he wondered if Christian Science practitioners just picked "out here and there the plums, leaving the real work to the real physicians."[40] Similar reasoning had led the editors of *JAMA* in 1901 to assert: "We might as well credit the reduced mortality of the last ten years to the influenza as to Eddyism."[41]

Christian Scientists protested such conclusions and repeatedly complained that the medical press reported only the deaths and rarely the healings under their treatment. To settle things once and for all, they called for a government-sponsored comparison of the success rates of various systems of healing, including Christian Science.[42] Apparently no agency conducted such a study, but W. M. Polk, dean of Cornell University's medical department, acknowledged that the questionable sources of Christian Science claims and not the results of their practice may have led the scientific community to view them "with less favor than others." He hastened to add that "medicine, ever seeking to push further and further its dominion over disease and death, has constantly before it new remedies—large sums in energy, education, intellect, and money are being expended the world over in this direction, and every measure of relief or cure is promptly heralded and tested by eager experts." Thus, despite their extraordinary origins, the claims of Christian Scientists finally had been "taken up and tested."[43]

The initial investigations by medical scientists exposed the apparently untrustworthy nature of the "evidence" collected by Christian Scientists and led some to conclude with H. V. Sweringen, M.D., of Fort Wayne, Indiana, that "the statistics that are at hand fully prove that those who have sought Christian Scientists professionally with any serious disease are now at rest in Abraham's bosom."[44] In fact, such anxieties arose about the effect of increased Christian Science and Divine Science activities on Denver death rates that the city's health commissioner in 1900 ordered separate records of the burial permits issued for those treated by such practitioners. He wanted to ensure that the public could "judge whether this increase in mortality is due to [un]improved sanitation or to 'Christian Science.'"[45]

To judge the efficacy of Christian Science treatment, physicians required their own statistical analyses of the cures claimed by Christian Scientists. In their view, without such investigations one had just anecdote, hearsay, and wishful thinking, which stood the Christian Scientists in no better position than all the other mind curers and quacks.[46] As early as 1886 J. P. Widney, professor of the principles and practice of medicine in the college of medicine of the University of Southern California, proposed a protocol for such studies. He believed that before mind cure can by the rules of scientific evidence claim to be established as a true science, [it] must be able to present cases of cure of organic disease, the existence of the disease and the fact of cure both being authenticated by competent observers—men who through study of disease are to be considered experts; or else it must be able to furnish tables, similarly authenticated, of cases of ordinary disease whose course has been sensibly modified and shortened by its power. As an illustration may be mentioned the tabulated reports given in medical literature showing the control which the cinchona salts have over malarial affections, mercury over secondary syphilis, iodide of potassium in the tertiary forms, or salicylic acid and its salts over inflammatory rheumatism. No such evidence has yet been furnished by the mind cure, and until it is furnished and authenticated beyond question, the verdict upon its claims as a true curative power in disease must be—Not proven.[47]

John B. Huber, M.D., at the time assistant to the chair of principles and practice of medicine in the New York University and Bellevue Hospital Medical College, conducted one of the first studies by a physician into the case histories of patients "cured" by Christian Science. Sponsored by *Medical News*, which published his findings in 1899, Huber examined twenty cases of alleged cure and sought, with little success, to interview numerous Christian Science officials, including the charismatic Augusta Stetson, then chief spokesperson of the First Church of Christ, Scientist, in New York City. Septimus J. Hanna, C.S.D., a former judge and at the time a chief legislative lobbyist for Christian Science, refused to answer the detailed questions Huber posed, in effect saying that the only way to fairly evaluate Christian Science was to become a Scientist.

Believing that cures of organic disease were the real test of medicine, Huber solicited only those supposedly cured through Science when "the structure of the organs" had been involved, and whenever possible he interviewed the patients' former physicians to learn of prior diagnoses and treatments. After Huber had completed his interviews and examinations, he determined that he had found no evidence that Christian Science had healed any of the subjects. Furthermore, in his view the movement was inherently dishonest and mercenary for not making an effort "to carry the blessings of 'Christian Science' among the poor." He allowed that "many Christian Scientists are undoubtedly good and worthy people" but concluded that the most important lesson to be learned from them is the "extraordinary readiness [of mankind] to be humbugged."[48]

Henry H. Goddard, A.M., fellow in psychology at Clark University, conducted an even more extensive survey and analysis of faith cures, including John Alexander Dowie's Divine Healing (over 1,600 "cures" examined) and Mental Science, of which he believed Christian Science was an offshoot. Over a two-year period Goddard conducted detailed interviews with patients who had been cured by practitioners and with those for whom the system had failed, surveyed *Science and Health*, and studied the life of Eddy. Although he found that Christian Science "unwillingly yielded its facts and philosophy," he completed his study and published his findings in a widely read and often cited 1899 article in the *American Journal of Psychology*.[49]

The following experience illustrates the difficulty Goddard and other scientists had in studying a movement whose language and metaphysics so contradicted their own. In an effort to construct personal profiles of his respondents, Goddard gave each the following instructions:

> Please answer the following questions in relation to your own personality, with great care. Age? Temperament? Disposition? Complexion? Married? Do you now, or did you as a child, choose or avoid responsibility? Did you, or do you, prefer solitude or companions? Were you precocious, backward or normal, in the matter of learning to write, walk or talk? What was your health in childhood?[50]

One respondent replied: "This paragraph is unanswerable from [a] Christian Science basis, since it [Christian Science] deals with mentality only, and recognizes physicality as the manifestation of mistaken, changing, human belief; having no fixed character of its own, and subject to constant correction."[51] In other words, since a person is really an idea of God and not a personality comprising body and mind, the kind of "biographical" data Goddard requested did not exist. In fact, a respondent who supplied such information would be perpetuating an error of belief in personality and hindering realization of the true self. Given such contradictory worldviews, was it any wonder that physicians and Scientists disagreed about what was good evidence for or against the others' system of healing? To discover whether patients got well under Christian Science treatment would have required more than the collection and collation of facts. It would have required an agreed-on worldview and an adjudication of the meanings of such concepts as human existence, sickness, and wellness. Most Christian Scientists and scientific physicians did not attempt that. Maybe Judge Hanna was right when he suggested that the only way one could really know if Christian Science worked was to become a Christian Scientist.

Although doctors thought Scientists' claim to Christian orthodoxy was folly, when they examined the scientific and medical claims of Christian Science their disdain knew no bounds. They found nothing but circular logic and speculative rationalism in *Science and Health*. The eclectic physician John M. Scudder spoke, perhaps, for most physicians when he concluded: "There is nothing of science in it—not even the definition of the dictionaries, 'facts collected and arranged.' There are no facts and no arrangement."[52]

The early studies of Huber and Goddard, the later work of Richard C. Cabot and Charles E. Humiston, and the experiences of a number of physicians suggested that Christian Science did have a demonstrable effect on functional disorders.[53] Different physicians listed different examples, but a composite list of the functional disorders cured included neurasthenia, nervous breakdown, hysteria, asthenopia, diffuse or shifting pains, joint aches or limitations, vicious digestion, faulty excretion, hypochondria, and psychoneuroses.[54] However, they also concluded that there existed no documentable case in which Christian Scientists had cured an organic disease. In such cases, at best they may have stimulated the body's own physical functions for short periods, but more often such supposed cures resulted from the self-limiting nature of many diseases.[55]

What Is Our Responsibility? Subsume and Curtail

A striking feature of the medical community's efforts to come to terms with mental and religious healing was its ambivalence about the subject—an ambivalence rooted in a widespread American confusion about the proper role for religion in general and Christianity in particular in a new scientific age. This confusion was especially noticeable in the case of Christian Science with its idiosyncratic medicoreligious nature, which repeatedly placed it at odds with both orthodox Christianity and scientific medicine. The medical community's efforts to struggle with the confusion often involved a debate over human nature.

An increasingly complex understanding of human anatomy and physiology grew to dominate the American medical view of human nature by the last third of the nineteenth century. Primarily understood in terms of various mechanical, chemical, and electrical models, for many physicians the body came to eclipse the soul in importance. But Christians still claimed to possess authoritative knowledge regarding human nature; and the will, with its seat the soul, continued to figure prominently in their view. Since presumably most American physicians remained at least cultural Christians, they were not yet ready to completely give up the spiritual nature of humans, even though they found themselves ever more comfortable with a mechanical understanding of self.

Religious healers thrived in Progressive Era America in part because they undermined this mechanical human by appealing to the still-strong popular belief that the ontological, if not the immediate, cause of disease lay in a fallen human nature. Healing therefore required some ministry to the broken soul of humanity and not just dosing of the material body. Christian Scientists carried this principle to an extreme when they radically extended the spiritual nature of humans to engulf not only the self but the whole universe, with the corollary ontological claim that not only did disease not exist, but neither did the body that supposedly became diseased.

When physicians, believing as many did about human nature and its relation to reality, analyzed the theory and practice of Christian Science, they often found it nonsensical. But they also saw it, however dimly, for what it truly was—a competing worldview that might threaten their new understanding of human nature and their growing authority over it. This threat presented itself through the whole array of mind or mental healers who flourished during the late nineteenth century. As we have seen, those curists had jettisoned many of the spiritual trappings of religious healing, but they continued, outside the medical mainstream, to practice a kind of suggestive therapeutics. Thus threatened by the competition of these mental and religious healers, including Christian Scientists, physicians worked to undermine their appeal by asserting that such healers could cure only functional, not organic disorders and that their healing techniques represented simply the latest manifestation of the well-known phenomenon of hypnotism. Granted, physicians soon supplemented "hypnotism" with terms drawn from a continually evolving psychological jargon. But by explaining the results of religious healing in their own psychological terms and thus subsuming them under their mechanical worldview, they took their first step to curtail mental and religious healers.[56]

Before about 1910, the medical community's key to understanding religious and mental healing lay in the physical processes of mind-body and mind-mind relationships. The physiologists, physicians, and psychologists who discussed mental healing and Christian Science in the medical press possessed a rich vocabulary to describe the faculties or processes whereby patients' minds affected their bodies during mental healing. These included faith, belief, the will, the placebo effect, great excitement, a sense of expectancy or anticipation, and the *vis medicatrix naturae* (nature's healing force).[57] Finally, however, before the growing American interest in Sigmund Freud's theories during the teens, they reduced them all to the principles in operation under hypnotism or the power of suggestion, the essential therapeutic elements of the psychotherapy movement.[58]

In just such a way Goddard concluded "that the curative principle in every one of the forms [of faith cure] is found in the influence of the mind of the patient on his body." He thought it interesting that each of the faith cure sects claimed the others were merely hypnotists, while in his view "*all* mental therapeutics is hypnotism, *i.e.*, it is suggestion" and the "law of suggestion is the fundamental truth underlying all of them." "In both hypnotism and Christian Science," he continued, "it is the *fixed idea* in the mind of the *patient*—placed there by the healer or operator, or suggested by a book or elaborated by the patient's own reasoning—that accomplishes the result through its tendency to 'generate its actuality.' "[59]

Physicians apparently were much less clear about the underlying physical mechanisms by which these psychological phenomena worked. By the turn of the century many thought some sort of electrical or chemical pathways between the brain and the body held the most promise as an explanation, but at least one Wisconsin physician believed the London Society for Psychological Research had demonstrated the existence of mental telepathy, which he thought accounted for the effect of a mind on a distant body, although he was not clear about how such imponderable forces might affect one's own mind.[60]

Those members of the medical community who sought to subsume and curtail mental and faith healing did not stop after they had translated them into their worldview. Some took the popularity of religious healing as an implicit criticism of the way they had been practicing medicine and proceeded to incorporate its attention to functional disorders into their own practices. These reformers believed that medicine had lost its way when it turned from considering the spiritual nature of human beings, and they believed physicians must rediscover their religious roots. But many more physicians held that the underlying mechanisms of religious healing must be discovered, translated into scientific language, and harnessed by medicine for the good of its patients. In either case, physicians worked to co-opt religious healing for their own healing practices by accenting the role of mind in maintaining health and conquering disease.

Physicians who decried the secularization of medicine believed that Americans in general had lost their sense of an immanent God who cared for their personal needs, and they blamed this in part on a clergy unwilling or unable to lend them succor. True religion, experiential and personal, had been replaced by stale dogma. They agreed with Robert M. Wenley, professor of philosophy at the University of Michigan, who argued in 1903 in Detroit's *Medical Age* that Christian Science drew its popularity from its appeal to a "deep-seated state of spiritual unrest" that had arisen among Americans "because dogma prevents religion from expressing itself in such a way as to face directly up to modern problems."[61] For this failure J. E. Engstad, a physician in Grand Forks, North Dakota, indicted the "arid waste of Protestant teachings" that "caused us to lose the idea of the belief of God's illimitable power and mercy," which Christian Science now restored with its doctrine that "we are of Christ."[62]

But it was not just the clergy's fault. While the American clergy had nurtured an often irrelevant Christianity that preached a present-day gospel devoid of the divine activities such as healing so notable in the apostolic church, physicians had stood idly by.[63] The problem was that many physicians no longer acknowledged the important connection between morality and health. As Hildegarde H. Longsdorf put it in 1894 to colleagues in the Medical Society of the State of Pennsylvania, Christian Science is "an old acquaintance with a new face, and its extraordinary progress has doubtless been due to the reactionary tendency of the times among a large class of orthodox people, from the scientific materialism, naturally growing out of the demonstrations going on here and abroad, as to the cause and prevention of disease."[64] D. A. Richardson, M.D., from Denver echoed these sentiments in his scathing critique of the secularization of medicine witnessed in his lifetime: "The regular profession has forgotten those greater, and by far the strongest, elements in human nature, namely, those of religion and morality, which require quite as much attention from a medical as from a physiological standpoint, and we are educating the masses by silence or example or both on these great subjects along retrograde lines instead of upward as we should."[65]

Given this, it was no surprise to Richardson that criminal abortion, fornication, and lasciviousness were all on the rise and leading to the spread of incurable diseases. Nonetheless he believed they could be prevented, but not by "medicine nor by rubber goods but truly, as the Christian Scientist tells us, through the upbuilding of sound christian character and the careful watchfulness of earnest christian parents over their children, and this can only be brought about by thorough plain speaking christian physicians, and I may add through such only."[66] Richardson believed that many Christian Science teachings were nonsense but that "their principles are right and worthy of all honor and acceptation if coupled to medical science, and, while disastrous, divorced as they now are from medical science, they nevertheless present to us a lesson we must learn."[67]

As doctors searched for lessons to be learned from the movement's growing popularity, few agreed with Richardson's view that physicians should become the moral as well as medical arbiters for society. Fewer, if any, agreed with the Scientists that Christian Science represented the dawning of a new age for medicine that would soon see all physicians become Christian Scientists. Scientific medicine, whether based on physiological processes or on germ theory, had grown to dominate the theoretical framework of most American doctors, and it left no room for the idealist metaphysics of Science. To have entertained such ideas would have meant succumbing to the sectarian impulses regular medicine had struggled against for decades. Therapeutics, however, proved a different matter.

By the 1880s American physicians had grown accustomed to thinking of medicine in implicitly positivist terms as inevitably progressing to ever more advanced therapeutic stages. "Ordered in this way," writes historian John Harley Warner, "the history of therapeutics functioned as a normative statement about future advancement and justified each author's particular program for reconstructing therapeutics by presenting it as part of a heroic historical progression."[68] Some predicted that the popularity of Christian Science, as well as other contemporary cults that worked through the power of suggestion, signaled the beginnings of a new stage in therapeutic history.

In 1898 the editors of *Medical News* submitted that the age of therapeutic confusion in which the country found itself was best seen as a transitional period, leading medicine to an advanced new age. Contrary to what most medical historians have argued, the editors believed Americans were experiencing a deepening distrust of medicine owing to the public's belated discovery of the "abandonment by medical men of the old doctrine of specifics for disease." However, the editors cautioned,

> there has been too much expression given, and by those least justified in giving it, to a lack of faith in remedial measures. Under the influence of the feeling of disrespect for what is called mere symptomatic treatment, by certain misguided, mostly young medical men, the good that may be done for symptoms is, as one of our distinguished old clinicians has recently said, left undone until symptoms have gotten beyond control; while if they had been conscientiously recognized and treated from the beginning much good might have been accomplished. The practice of medicine is an art, not a science, and is likely to continue so for many years yet to come.[69]

Medical News was not alone in thinking that maybe a new therapeutic day had dawned. By 1900 *JAMA* editorialized that "we have, perhaps, entered on an era of 'suggestive therapeutics,'" characterized in the present century by "Perkinism, Mesmerism, Keelyism, Eddyism, faith-cure, [and] osteopathy."[70] Ten years later, however, on the occasion of Eddy's death, *JAMA* reconsidered the role of Christian Science. Still acknowledging that it was not strange that idealism's "easy short cut out of theoretical materialism should be made the basis of a new cult," the editors now suspected that "the cult will not continue long in its present condition" and predicted, "judging from similar movements of the past, that it has reached its high-water mark."[71]

While not everyone believed that Christian Science symbolized the wave of the future, physicians could not deny that it proved attractive to many Americans and that cures of functional disease often followed its practice. But could they learn from its popularity? Some thought its reputation arose as a backlash against the way physicians practiced medicine, while others believed it prospered because it filled a void in contemporary therapeutics. Among the former group were those who believed that all religious healing (especially Christian Science) prospered because the new emphases on physiology and germs in scientific medicine and an overdependence on surgery had led physicians to ignore the psychic dimensions of human well-being. St. Louis physician Alexander B. Shaw, writing in 1889, concluded "that medical men, even of the highest attainments, are by the allurement of the possibility of discovering a physical cause for every ailment, and this cause a bacterial germ, and a germ[i]cide to remove each cause . . . have failed for many years to pay sufficient attention to what may be aptly called psychical disorders of the body, or psychical conditions engendering functional derangements, or functional disturbances produced by psychical states."[72]

Other physicians questioned the profession's use of surgery as well.[73] P. Maxwell Foshay of Cleveland complained in 1900 about the "slipshod ways of practicing medicine" that often led mediocre doctors "to the belief that they are competent to make a surgical diagnosis and to operate" with the result that they "are responsible for a very great deal of the quackery and voodooism that steadily flourishes in our midst."[74] Ten years later D. W. Harrington, president of the Brainerd Medical Society in Milwaukee, Wisconsin, echoed these concerns and suggested a financial explanation. He worried that "the profession has been so interested in morbid anatomy and surgery during the last score of years that it has neglected some less remunerative fields of practice and this is in no small measure responsible for the spread of drugless healing among those who have no knowledge of disease."[75]

Despite these criticisms from the medical community, doubtless most physicians believed it would be foolish to give up drugs and surgery, for without them they would be as defenseless against organic disease as the Christian Scientists who blindly ignored the "symptoms of serious disease" until it was "too late to apply rational methods of treatment."[76] True, not all agreed with Omaha, Nebraska, physician J. M. Aikin that "the success attending any physician's practice is due in greater measure to the mental impress he stamps upon his patients than to the drugs he gives."[77] Nonetheless, many conceded that, as one Cincinnati physician put it, "the average of us physicians use mental potencies far too little, being inclined to be idolaters of drugs."[78] But how were physicians to remedy this state of affairs?

Whether or not the new vogue of germ theory or the greed of surgeons was to blame, many practitioners persisted in their belief that despite the advances in medicine, medical therapeutics lay in disarray and that it was not surprising that new sects arose to feed the needs of an anxious public. A. W. Abbott, president of the Minnesota State Medical Society, voiced just those worries when he addressed his colleagues in 1893:

> Sects can only spring up in doubtful soil. . . . So any school of medicine will become preeminent and above criticism as it approaches the standard of an exact science. Empiricism was good when we had nothing better, but the present is an age of positive proofs; nothing short of demonstration satisfies. There are no schools of anatomy, chemistry, physiology, and we may now say of surgery. These have become sciences. The field of therapeutic medicine is the Donnybrook fair of our cult. The regular school is founded upon no single theory. . . . But until diphtheria, the exanthemata, malaria, tuberculosis, in fact all non-surgical diseases are absolutely understood, and an infallible cure for the early stages, at least, found, the sick man will still insist that one guesser is as good as another.[79]

A few physicians viewed the popularity of Christian Science as less a backlash against medicine than the filling of a void left by current regular therapeutics. As a remedy they suggested an updated environmentalist model of therapeutics that would require physicians to pay closer attention to the social and cultural circumstances surrounding their patients. Hildegarde H. Longsdorf believed that Christian Science had grown as a response to "our modern high pressure civilization" that had rendered the public more willing "to believe what it cannot explain, and revere what it cannot comprehend."[80] Eclectic physician R. L. Thomas agreed and observed in 1899:

> We are living in a fast age; competition is great and the delicate nervous system is keyed to the highest tension and our imaginations are running riot. A large percentage of cases are nine-tenths imaginary, and one very important factor in the cure is in impressing the *mind* with the fact that they positively will get well.
>
> Here is where the Christian scientist, the Dowieite, and the "Faith Healer," succeed in their cures. The patient believes *implicitly* he will be healed; a great impression is made on the mind, the tension is removed, and the individual complains no more and is cured. This is quite common in persons of vivid imagination, and in cases where there is but little pathology.[81]

Other early efforts to incorporate the truth behind Christian Science into medical practice called for physicians to reinstill confidence in their patients by reasserting their authority over disease.[82] For L. C. Mitchell this meant promising more patients a cure; for E. W. Mitchell it meant that "the true physician must be a true philosopher," liberally educated, who recognizes "the importance of reading more than our medical books"; and for Alexander B. Shaw it meant that when patients asked what drugs or dosages were being administered and how they worked physicians would "courteously, yet positively indicate that we are posing as practitioners and not as teachers of the healing art. That our services as physicians are at their disposal, but that we have neither the time nor the inclination to make half-way doctors of them."[83] These early lessons learned by physicians from Christian Science seemed to lead back to a recovery of the past glories of halcyon years by reasserting lost beliefs and practices. In contrast, a small group of Boston physicians in the early 1900s creatively explored the territorial boundaries evolving between clergy and physician in order to map out the terrain of psychotherapeutics.

Amplifying the founding efforts of Episcopal clergyman Elwood Worcester (1862–1940) and others, James Jackson Putnam, Richard C. Cabot, Isador H. Coriat, and Joseph H. Pratt worked to combine religious counseling and medical psychotherapeutics within the Emmanuel Movement. Although the specific circumstances for their attachment to the movement varied, each of these physicians objected in some way to the generally practiced therapeutics of his day, especially as it related to the role of mind in health, and drew the conclusion from the popularity of religious and mental healing that physicians needed to experiment with new approaches.[84]

Cabot, who repeatedly irked his colleagues with his public criticisms of the medical profession, commented that simply to identify "cases of diphtheria refused antitoxine under the fatuous advice of the Christian Scientist is not to prove that mind cure is bad. They kill people, of course, now and then; so do we. But nevertheless they are useful in the long run just as we are."[85] In his view the "mind curist does a good piece of public service by pressing upon us, in his own peculiar jargon, the truth that disease and all that reminds us of it, including the doctor, should be forgotten and put out of sight as far and as fast as possible, and that health and the interests of health are the only proper contents for every mind in the community except the doctor's."[86]

Cabot and Putnam, however, soon withdrew from association with the movement and joined a growing number of physicians troubled by the "amateur" encroachment into psychotherapeutics by ministers untrained in science or medicine.[87] Although the movement's marriage of medicine and ministry continued to receive the support of some American physicians, most seemed to share the view of W. A. Jones, M.D., editor of the *Journal of the Minnesota State Medical Association and the Northwestern Lancet*, that "the clergyman has already more than he can do to keep the morals of the public free from contamination. It may not be wise to attempt a mixture of the two professions."[88] Ironically, the movement may have been for many Americans the bridge that carried them from the supernaturalism of religious healing to the naturalism of modern psychology.[89]

The more lasting impact of Christian Science on psychotherapy lay in the small part it played as a catalyst to stimulate some American physicians to take mental health seriously and to adopt psychoanalysis. The popularity of Christian Science may have prodded medical educators to incorporate training in psychotherapy into the curriculum for medical students. Although pioneer psychotherapists such as James Jackson Putnam had argued early in the century that "psychology was as important for physicians as physics and chemistry," by 1910 such a view had won only grudging support, and then mainly only at university-connected hospitals in some eastern states.[90] As late as 1914, leading American neuropsychiatrist William A. White concluded that "in the matter of the ills of the mind, physicians today are, on the whole, quite unable to cope with the problems involved." He believed physicians needed to be educated in the care of the mind, otherwise these patients become "one of the greatest sources of revenue of the charlatan" and faith curist. Each "gets what the physician cannot deal with."[91] Behind the leadership of advocates like White, twenty years later medical education in the treatment of mental illness had improved significantly, possibly in part to co-opt for physicians those patients who patronized Christian Scientists and other "charlatans" for ills that physicians had often ignored. But in that process, rather than expanding instruction in hypnotism and suggestive therapy, which had grown indistinguishable to many physicians from what Christian Scientists and other mental or religious healers did, they adopted psychoanalysis, which, at least in their minds, bore no resemblance to Christian Science.[92]

Progressive Era American physicians persistently adopted a strategy against Christian Scientists of attack and sublimation. They worked to make Christian Scientists' teachings appear ridiculous and their healings seem fraudulent while hesitantly adapting and absorbing into their own practices the emphasis on mind of mental and religious healers. As we will see, by working legally to confine the practice of all healing to licensed medical doctors, physicians mounted their ultimate campaign to curtail Christian Science and place it under their control.[93]

Therapeutic Choice or Religious Liberty?

Inasmuch as the medical profession is in a better position to know the truth about health and disease than any other body of men and women, inasmuch as the medical profession has from time immemorial devoted all its efforts to conserving the health of the community and to stamping out disease, and inasmuch as our own personal interests are jeopardised by the medical activities of the Eddy Healers—it is the business and duty of the medical profession to wage incessant war on the Eddyites, to watch over the activities of legislatures, to keep a wary eye on the conduct of our courts and to prosecute every breach of the law.
—Samuel A. Tannenbaum, M.D., 1916

If the practice of medicine were universally successful, and if it could be established that no other system heals the sick, there might be some excuse for such legislation as interested parties have tried to obtain in several states, but the evidence is so overwhelmingly on the other side that it discloses the oppressive nature of the legislation and the selfish purpose of those demanding it.
—Willard S. Mattox, C.S.B., 1903

"Long before the hour set for the trial, 2 o'clock, the courtroom was crowded, the majority of the spectators being fashionably-attired women who had probably never seen the inside of a Police Court before. They were Christian Scientists, and some had come a long distance to hear the trial, as the case has attracted much attention." So began the *Cincinnati Enquirer*'s report of the 1898 trial of forty-year-old practitioner Harriet Ohio Evans, "found guilty by a jury in the Police Court last evening [8 December] of unlawfully practicing medicine."[1]

Mrs. Evans, a dressmaker for twenty years, had become a believer in

Christian Science because she had been "thoroughly healed in Lima, Ohio, in 1889 by it." She had attended school for eight years, until her sixteenth birthday, but she had never attended college, seminary, or medical school. Then in 1891, doubtless eager to learn more about her new faith and receive the instruction increasingly necessary for a reputable practitioner, Evans had twice attended the Christian Science Primary course of Hannibal P. Smith and Etta Beatty in Philadelphia. Since 26 January 1894 Evans had resided with the wife and family of Thomas McDowell, the man whose death from typhoid fever on 13 November 1898 had brought her before the police court of Cincinnati.

Unfortunately for Evans, the autumn of 1898 was not a time when Christian Science was popular in Cincinnati. During October and November the Lutheran Rev. J. S. Simon had preached a series of sermons against "Christian Science as a Religion," in which he ridiculed its doctrines, "challenged its claim to being the true apostolic religion," and refuted the claim of Mary Baker Eddy ("the high priestess of Christian Science") "to a final revelation" of divine truth. On the eve of Evans's arrest the *Enquirer* carried a lengthy and unflattering report of Maine practitioner "George Wheeler, a long-haired, grewsome individual, [who] posed as a purveyor of faith, and even went so far as to hang out his shingle advertising himself as [a] regular[,] practicing Christian Scientist." Adding to these immediate tensions of 1898, according to the *New York Times*, was the fact that "ever since the [Ohio] Legislature passed a law in 1896 compelling all physicians to pass an examination and register, a vigorous crusade has been instituted against Christian Scientists who take it upon themselves to cure sick persons."[2] Therefore, when Evans's landlord died while under her care the public outcry could not have been completely unexpected.

Even before McDowell's death a public dispute had arisen over his care. Soon after typhoid fever had driven McDowell to his bed in late October, he "had been prescribed for by a physician," but "he became rapidly worse."[3] Ten or twelve days later, according to the *Christian Science Journal*, McDowell's wife, a Christian Scientist, enlisted Christian Science treatments for her husband from the family's boarder, Harriet Evans, who may have treated him and his wife on previous occasions.[4] Evans immediately began treatment on 4 November and continued daily for six days, but McDowell's condition steadily declined.[5] Alarmed by these developments, Lulu—McDowell's daughter and his wife's step-daughter—summoned Dr. Charles F. Wocher on 10 November "to attend him and [he] prescribed medicine and the usual baths." However, "After the doctor had made several calls he learned that the treatment was not being followed. The medicine was not being given to the patient and he was not subjected to the baths." We may never know which treatment Mr. McDowell preferred or if he wanted both, because the testimony of Lulu McDowell conflicted with that of Mrs. McDowell and Harriet Evans. At the coroner's inquest Evans guardedly asserted that "she [had] administered to McDowell the faith cure by request, but had not said that he should not take the medicine given by the regular physician." However, Lulu testified that "I never heard my father ask for her [Evans]," and the *Cincinnati Enquirer* quoted a "well-known believer in Christian science" as saying that "a practitioner that is versed in Christian science would never attempt to force herself on a patient like Miss Evans did in the McDowell case." Enough confusion surrounded the question whether Mr. McDowell had requested Christian Science treatment that on the eve of his death "a brother of Mrs. McDowell, and two women, Christian Scientists, went to the house and raised McDowell up in bed only a few hours before his death, and got his signature to a paper exculpating Miss Evans."

When Cincinnati Health Officer W. A. Tenney learned of the conflict over treatment and the involvement of a Christian Science practitioner, he communicated with Dr. Charles A. L. Reed of Cincinnati, a member of the Ohio State Board of Medical Registration and Examination and later the president of the American Medical Association (1900–1901), who at once notified Dr. Frank Winders of Columbus, secretary of the state medical board. Winders arrived in Cincinnati during the afternoon of 11 November and, accompanied by Reed, called at the police court clerk's office, where the clerk issued a warrant for the arrest of Harriet Evans. Reed announced, "We intend to prosecute this case to a finish. We shall drive these people [Christian Scientists] out of practice. Every time we hear of them attending a case, we shall make an arrest."

That proved to be true, at least for the short term, for only two weeks later in the same Cincinnati police court, Christian Scientist Allie Putnam was found guilty of practicing medicine without a license and fined $100 and court costs. But in this case the state board of health had not waited to "hear of them attending a case," but had sent Carey B. McClelland to seek treatment.

[McClelland] testified that he had called at the offices of the Christian Science Healing and Literature rooms in the Odd Fellows' Building, and represented that he was ill. There he found Miss Putnam, who asked him numerous questions as to his complaint. After fixing the fee for treatment she placed a handkerchief over his eyes and appeared to be in deep thought or silent prayer. McClelland called on her again at her residence, at 119 West Seventh street, on the 6th of this month [December]. She went through the same maneuvers as before. He subsequently called on two different occasions and received similar treatment for his supposed ills.[6]

Two days after Evans's arrest but before the case against her for practicing medicine without a license had proceeded to trial, McDowell died. Ralph Westfall, attorney for the State Board of Medical Registration and Examination, "stated that the board would endeavor to secure Mrs. Evans's conviction under the laws regulating the practice of medicine in this state, and in case this failed she would be held on the charge of manslaughter."

The testimony taken during the trial revealed much more than the particulars surrounding the death of Thomas McDowell. It opened to public view the healing ritual of Christian Science as seen through the eyes of both believers (Harriet Evans and Mrs. McDowell) and skeptics (Lulu McDowell and Elizabeth McDowell [his other daughter]).

When asked how she had treated McDowell, Evans answered: "What I did chiefly consisted in reciting the Lord's prayer, for by it we know the sick are healed. There was no definite time as to the length of the time I prayed. Sometimes 15 minutes and sometimes 30 minutes. When Mr. McDowell asked me to help him he said he would give the money to me instead of paying it out for medicine." When asked how much she charged for prayers, "she replied that Christian Scientists charged for their time" and "then said that Christian Scientists could heal a broken leg without bandage or surgery, and that it was their belief that there was no transmission of contagious diseases."

Mrs. McDowell, the widow of the deceased, reported that her daughters "were excluded during the treatment, as the rule is to be alone with the patient." Elizabeth McDowell "testified that she had seen her father get out of bed while he was sick and walk across the room supported by Miss Evans. The latter recited prayers and her father repeated them."

When prosecuting attorney Lueders asked Lulu McDowell, "How did Miss Evans treat him?" she testified:

> She prayed for him. Sometimes she would sit beside him with her head resting in his [her?] hand as though in deep thought. . . . Miss Evans got up at night sometimes, to wait on my father. On Sunday he was delirious and said things that made us laugh[.] Miss Evans did nothing when he was that way, but would laugh it over. I never heard my father ask for her. I had several quarrels with her about the case. The first time we quarreled was when she wanted to give him plums. I said that he ought not to have them, as plums were not good for persons who had typhoid fever, but she said that they would not hurt, as he had diphtheria. He did not get the plums, and that evening Miss Evans and I had another argument. Another time I said while Miss Evans and Mrs. McDowell (her mother) were in the room that if they didn't have a doctor for papa they would be arrested. Miss Evans said that she would be proud to be arrested to make known their teachings.

Lueders asked: "Did you ever speak to Miss Evans about Christian Science?" Lulu responded:

> I did. . . . I told her he could not be cured [by] that, as he was too sick to be trifled with. She said: "I know better," and stamping her foot, called me a liar twice. She said papa [w]ould die if there were too many doctors in the case and that I was killing him because I wanted one called. Mrs. McDowell is my stepmother. She is a Christian Scientist. The only book I saw Miss Evans have was one called "Science and Health."

Prosecutor Lueders stoked an already overheated atmosphere when he concluded his final argument to the jury by charging that "Miss Evans's methods and hoodoo practice among colored people were in the same class" and that in my "opinion the charge should have been manslaughter."

Although the jury returned a guilty verdict after only twenty minutes and Evans was fined $100 and court costs, the decision did not stand the challenge of appeal. On 30 December 1898 Judge J. Hollister of the Hamilton County Court of Common Pleas concluded that the state statute in question, which precluded unlicensed persons from practicing medicine or surgery, "was intended to embrace two classes of persons, those who use drugs or medicines, and those who, in cases of fractures, wounds, etc., use and must use some other agency, effective in that class of injuries." Since "there is nothing in the information to show that the 'system of christian science' is either drug, medicine or other agency of the kind described," then "this act does not make the practice of that system any offense." He concluded: "The court is of the opinion that the law in question does not include such acts as the defendant is charged with having committed, and for all the reasons given above the judgment of conviction is reversed."[7] For Christian Scientists it was "ANOTHER VALUABLE JUDICIAL DECISION."[8]

Apologists for Christian Scientists have often painted a rosy picture of the believer as simply following her conscience, while detractors have dramatized the danger to the nation's public health in the practice of all unorthodox healing, including religious healing. As the experience of Harriet Evans illustrates, neither approach captures the complex nature of the interaction between believer and doubter that such trials often revealed. For neither party was it simply a matter of conscience.

In the first place, Christian Scientists did not just quietly and conscientiously practice their own variety of Christianity; they actively proselytized for their beliefs and healing practices, which they believed constituted a restoration of the true message of Jesus and his apostles. Although they rarely pursued open conflict with other healers or with public health officials, neither did they timidly avoid confrontation, for when it arose the public could learn more about Christian Science. In 1895 the editor of the *Christian Science Journal* wrote that

> our enemies, are proving to be our friends. They cannot openly attack Christian Science without advertising it, and they cannot advertise it without causing some to investigate it. . . . Let the M.D.'s and the D.D.'s, the little guns and the big guns, keep up their fusil[l]ade, for they are building for Truth, and their imaginary wrath is being made daily to serve God. Let those who are being persecuted for Truth's sake, who are being arrested and taken before the courts, take courage, stand valiantly by their colors, and know that they are but instruments in the hands of God for the more rapid spread of Truth.[9]

Harriet Evans exemplified this vision when she testified "that she would be proud to be arrested to make known their [Christian Science] teachings."

But the testimony at Evans's trial highlighted more than the public conflict at center stage; it briefly pulled aside the curtain from the private conflicts among patient, practitioner, and patient's family that often complicated treatments in families of mixed faiths. Many Christian Scientists were the only believers in their homes. Could very ill, nonbelieving members of the family be treated without their consent? Could a practitioner treat a patient who was under the care of a physician?

Although we will never know for certain whether McDowell requested Christian Science treatment, it is clear that his wife and daughter vehemently disagreed about its wisdom, and the issue of patient consent greatly concerned the Christian Science community. At least one practitioner believed Evans had overstepped her bounds. As we have seen, she commented to a newspaper reporter, "A practitioner that is versed in Christian science would never attempt to force herself on a patient like Miss Evans did in the McDowell case." Furthermore, the Christian Science community sought to protect itself and exonerate Evans by getting McDowell's signature "only a few hours before his death." Failing to gain McDowell's consent would have been both a dangerous theological error and a potentially troublesome public relations misstep. In the first place, Eddy had argued for decades that treating patients without their consent constituted mental malpractice (the use of "mesmerism" or "malicious animal magnetism"), something strictly forbidden in Science. Second, for Scientists, who repeatedly argued that Americans should have therapeutic freedom of choice and who claimed that any legislation to the contrary represented an invasion of personal rights, consent for treatment was critical to their credibility.

The issue of consent created tensions, but Evans's unwillingness to administer treatment while McDowell was under the care of a physician ignited its own explosion. As we have already seen, Lulu testified, "I told her he could not be cured [by] that, as he was too sick to be trifled with. She said: 'I know better,' and stamping her foot, called me a liar twice. She said papa [w]ould die if there were too many doctors in the case and that I was killing him because I wanted one called." Miss Evans's zeal for Christian Science treatment and her vehement rejection of the therapies of physicians typified practitioners' beliefs about the incompatibility of medicine and Science. But to many nonbelievers it appeared to be unfounded dogmatism. Further confusing matters, Lulu testified that she had thought her father had typhoid fever, but Evans said he had diphtheria. This offhand diagnosis of a disease, whose existence Christian Scientists denied, typified the informal diagnosis that often preceded Christian Science treatment. But to unbelievers it seemed inconsistent with Christian Scientists' claims to reject medicine and only strengthened the public's doubts and antagonism.

Although many doubters of Christian Science conscientiously warned Americans of its dangers for the nation's public health, other factors of equal importance influenced their activities. One of the attractions of Christian Science healing was the way it unified the process for attaining physical health and spiritual well-being; but this also laid it open to attack by both religious and medical critics. The clergy of most well-established denominations self-interestedly derided the restorationist claims of Christian Science and defended their own theologies of sickness, suffering, and spiritual healing.[10] Physicians also were embroiled in their own efforts to assert their expertise over matters of health and establish the legal basis to perpetuate that authority. When the Reverend J. S. Simon ridiculed the doctrines of the "high priestess of Christian Science" and Prosecutor Lueders compared Christian Science healing to the "hoodoo practice of colored people," they were doing more than protecting the public's well-being; they were trying to form the public's image of Christian Science and of religious healing in general to suit their own ends. By holding it up to public ridicule they were seeking to deny it a fair hearing and strengthen their own authority.

Whereas scholars have established that among Americans there have perennially been those who held to a *belief* in religious healing, it is also clear that often those same Americans have been ambivalent about the *practice* of religious healing.[11] On the one hand, the United States has generally tolerated and at times even encouraged adults' exercising the freedoms of religion and therapeutic choice by ministering to their own souls and bodies as they see fit. On the other hand, when orthodox healers have convinced a majority of Americans that orthodox therapies are demonstrably more efficacious than others, they have often sought to impose controls on the practice of unorthodox healing, including religious healing. Such controls have been justified on the grounds that unorthodox healing endangers a supposedly gullible or desperate public, especially children, but the controls have run the risk of infringing on individuals' autonomy over their bodies or their right to believe and practice their religion.

Because religious healing lies at the heart of Christian Science's reason for existence and constitutes one of its most visible public activities, Christian Science practitioners since the late 1880s have been buffeted by this American ambivalence. Despite the innovative use of lecturers, dispensaries, and Reading Rooms, the shock troops of Progressive Era Christian Science remained the practitioners, who found it increasingly difficult to legitimate their healing practices in the face of the strident professional authority asserted by medical doctors. The ways physicians, Christian Science practitioners, and the public juggled individual liberties and community safety can be illuminated by examining each group's efforts to accommodate the demands of medical licensing laws and public health ordinances requiring vaccination and quarantine.[12] The public arenas of accommodation or confrontation were often courtrooms. Between 1887 and the death of Mary Baker Eddy in 1915, more than thirty criminal or misdemeanor cases involving charges against Christian Science healers went to trial, and scores of others went before coroners and grand juries that failed to indict (see appendix).[13] Practitioners found themselves cited, indicted, and tried for practicing medicine without a license, breaking public health ordinances (for example, those requiring immunization or the reporting of contagious disease), or even committing manslaughter or murder, despite their protestations that they were just practicing their religion or offering individuals a therapeutic alternative. This chapter explores the legislative and court battles between physicians and Christian Scientists over medical licensing, the meaning of medical practice, and the supposed right of Americans to therapeutic choice.

Medical Licensing Laws and Christian Science

The late 1880s and the 1890s witnessed an expansion of state legislation intended to tighten the requirements for medical licensure.[14] The reforms instituted, which varied from state to state, included establishing state boards, requiring physicians to pass an examination and register with the state to practice, and mandating more rigorous medical school curricula.[15] Scholars have ventured a wide range of explanations for this flurry of lawmaking. Some have seen it as a response to the educational reforms in scientific medicine begun a decade or so earlier that coincided with growing cooperation among regulars, homeopaths, and eclectics.[16] Others, including the Christian Scientists enmeshed by the new laws, viewed it as an effort by physicians to achieve a competitive advantage or even a legal monopoly over the practice of medicine.[17] Still others have seen the efforts of physicians, like those of other occupations, as reflections of wider Progressive Era social efforts to "bring public affairs under the sway of certified expertise," especially in matters of public health.[18]

Whatever the true motivations underlying these state efforts to license physicians, the unanimous decision by the United States Supreme Court in *Dent v. West Virginia* (1889) affirmed the state's police power in these matters. West Virginia's board of health had denied a medical license to Frank Dent, a graduate of the American Medical Eclectic College of Cincinnati, because it claimed that "said college did not come under the word 'reputable,' as defined by said board of health." Dent argued that the state's decision had deprived him of the right to practice his profession, a right protected under the Fourteenth Amendment's declaration that "no state shall deprive any person of life, liberty, or property without due process of law." Justice Stephen Field, speaking for the Court, rejected Dent's claim by balancing the right "of every citizen of the United States to follow any lawful calling, business, or profession he may choose" and "the power of the state to provide for the general welfare of its people" by establishing occupational standards. "The nature and extent of the qualifications required," which undoubtedly would change as knowledge increased, continued Field, "must depend primarily upon the judgment of the state as to their necessity." "No one has a right to practice medicine," concluded Field, "without having the necessary qualifications of learning and skill; and the statute only requires that whoever assumes, by offering to the community his services as a physician, that he possesses such learning and skill, shall present evidence of it by a certificate or license from a body designated by the state as competent to judge of his qualifications."[19]

Both economic and political populists joined the publics opposed to these legislative efforts to regulate the practice of medicine. While some objected to medical licenses because they contradicted the principles of economic laissez-faire that they felt should prevail in America, others believed the state should not legislate one's supposed freedom to choose one's own healer, no matter what the healer's "established" competence. But both groups expressed a distrust of government and expert elites' authority to legislate in their best interests on matters of health. Some thought government was corrupt and therefore untrustworthy, but many more simply thought government was untrustworthy because it comprised elites concerned with their own self-interest and not with the interests of common people like themselves. It appears that what these opponents of licensing really distrusted was the authority of the state or its appointed experts to determine the nature and cause of disease and to thereby impinge on their "sacred right" to choose their own medicine and doctor. Among such opponents of medical licensing were the Christian Scientists, who opposed licensing because it intruded into their practices. But was Christian Science healing the practice of medicine? Initially, for most members of the medical community this question required a simple yes or no answer; but for most Christian Scientists the answer was yes and more.

Christian Scientists believed they practiced medicine, but in their minds that represented only half of what their system of healing accomplished for a sin-sick world. Christian Scientists practiced a unified "system of medicine" and "system of ethics."[20] The all-encompassing nature of this healing system became apparent in the first issue of the *Journal of Christian Science*, published in 1884, when Eddy explained the advantages of her "system of healing, over the ordinary methods of healing disease":

> 1st. It does away with all material medicines, and recognizes the fact that the antidote for all sickness, as well as sin, may be found in mind, as is also the cause of all "the ills that flesh is heir to." 2d. It is more effectual than drugs, and cures where they fail; thus proving that metaphysics is above physics. 3d. A person who has been healed by Christian Science is not only healed of their beliefs of disease, but they are improved morally at the same time; the body is governed by mind, and mind must be improved before it can govern the body harmoniously. The body is but thought made manifest.[21]

Note that in each case Eddy first highlighted the supposed superiority of her system as an alternative to medicine; only then did she accent its value-added effects on sin and immorality. Although there appeared to be a symbiotic relationship between sin and sickness in Christian Science whereby demonstrating over either could lead to eradicating the other, the practical reality was that most people initially came to Science for a cure from sickness. The early issues of the *Journal* reflected this pattern over and over; from the advertisements for Eddy's classes to the answers given to the questions of inquisitive correspondents, the appeal of medical practice and treatment stood out.

Understandably, Christian Scientists took exception to physicians' efforts to regulate the practice of medicine and thereby infringe on their religion of physical and spiritual healing. In an anonymous 1886 attack on these attempts, an author in the *Christian Science Journal* complained that "our religious freedom is not to be curtailed in the interest of a hoary and decrepi[t] theory of physic, erected into a craft as a source of wealth and social respectability."[22] However, the Scientists did more than simply complain among themselves about the monopolistic efforts of doctors; they also kept themselves apprised of pending state medical legislation. Although their numbers were still small, during the late 1880s they spoke publicly in opposition to it, at least in Rhode Island and Massachusetts.[23] Rather than viewing themselves as a minor irritation to physicians, they believed their "truth" had struck a nerve so raw that, in the words of John F. Linscott, C.S.D., "the pulpit and the schools of medicine are now arrayed in a united effort against us, because the destruction of their craft is threatened."[24] Despite such optimistic views of their impact, however, these efforts initially proved of little effect and led the Christian Scientists to feel persecuted at least as often as they felt like a David to the medical establishment's Goliath.

At first Christian Scientists' initiative to protect their privilege to practice went largely unnoticed by members of the medical community and other influential citizens. When it caught their attention in the late 1890s, however, it confirmed physicians' belief that Christian Science was the practice of medicine, albeit quackery, and that the medical community would have to make serious efforts to ensure that medical practice laws included Christian Scientists. C. H. Hughes of St. Louis, editor of the *Alienist and Neurologist,* warned his colleagues in 1899 that "the Christian Science craze is fast becoming an epidemic delusion. Even the courts have decided in its favor as a variety of healing entitled to legal protection."[25] On the occasion of the recent graduation of Eddy's final class, taught at Concord, New Hampshire, 20–21 November 1898, the *New York Times* quoted a Christian Science press release to the effect that Eddy had "just completed an examination of a class of about seventy of the active workers in Christian Science mind-healing to confer on them the degrees of the Massachusetts Metaphysical College as healers and teachers of this system of medicine." The *Times* editor wondered, "Why is it not possible to suppress these murderous fanatics in this country? Mrs. EDDY's statement shows that she professes to have established a 'system of medicine.' . . . Before they are allowed to practice as 'healers' they ought in every case to be examined by the boards constituted by law to inquire into the character and qualifications of physicians."[26]

Christian Science Practitioners and the Courts

If in 1898 these "murderous fanatics" still practiced unsuppressed, it could not have been for lack of effort on the part of physicians and public health authorities. Since the arrest of Christian Science practitioner Charlotte Eddy Post in McGregor, Iowa, in early 1887 for practicing medicine without a license, at least four Scientists, including Post, had been brought to trial shortly after the passage of state laws requiring those who practiced medicine to register or acquire a certificate to practice in the state (see table 1). The table indicates in boldface the years of trials that occurred soon after medical licensing legislation passed in their states. There appears to have been no such correlation between new statewide legislation and a trial in three other cases, and in the case of Agnes Chester it appears that Michigan physicians used the publicity surrounding her trial to promote the need for more restrictive licensing of those who practiced medicine.[27] The trials of Ezra M. Buswell in Nebraska (1893–94) and Walter E. Mylod in Rhode Island (1898) both ascended to their state supreme courts and established important legal precedents. Although courts outside of these states' jurisdictions were not bound by their decisions, attorneys and courts in many other states cited them as precedent for their own arguments. The most important variable among the various states in determining the guilt of Christian Scientists proved to be a state's legislative definition of "medicine" or the "practice of medicine."

Ezra M. Buswell and his wife Elizabeth, among the first Christian Scientists west of the Missouri River, organized a Christian Science church in 1891 in Beatrice, Nebraska, which by 1893 had about eighty-six members.[28] In February 1893 a Gage County grand jury indicted Ezra for "Practicing Medicine and Surgery and Professing to Heal Physical

Table 1. Medical Licensing Laws and the Trials of Practitioners

State	Initial Postbellum Law	Single Board to Review Diplomas	Registration or Certificate Required	Date Charge Filed	Name
Iowa	**1886**	**1886**	**1886**	**1887**	Post
Nebraska	1881	1891	1881/1891	1887[a]	Bunnell
Nebraska	1881	**1891**	1881/**1891**	**1893**	Buswell
Kansas	1870	1901	1901	1895[b]	Graybill
Michigan	1883	**1899**	**1899**	**1896**[c]	Chester
Rhode Island	**1895**	**1895**	**1895**	**1897**	Mylod and Anthony
Ohio	1881	**1896**	**1896**	**1898**	Evans and Putnam
Nebraska	1881	1891	1891	1899[d]	Hammett
Minnesota	1883	1887	1883	1899[e]	Brookins and Meyer
Wisconsin	1867	1897	**1899**	**1900**	Arries and Nichols
Ohio	1881	**1896**	**1896**	**1903**[f]	Marble
New York	1874	**1907**	1880	**1911**	Cole

Note: **Bold** dates indicate a correlation between laws and trials. With the exception of the 1881 registration date for Nebraska, the dates of medical licensing laws taken from Samuel L. Baker, "Physician Licensure Laws in the United States, 1865–1915," *Journal of the History of Medicine* 39 (1984): 173–97. For Nebraska's 1881 registration law see Guy A. Brown, comp., *The Compiled Statutes of the State of Nebraska, Comprising All Laws of a General Nature in Force July 1, 1881* (Omaha, 1891), 347–49.

[a]Local physicians charged that Bunnell failed to register as the 1881 law demanded.

[b]This involved a municipal ordinance regarding occupational licenses.

[c]Physicians used this as a "test case" to raise public awareness of the need for more restrictive legislation. See "A Peculiar Case," *Christian Science Journal* 14 (1896): 34–35.

[d]Local physicians brought the complaint. Some evidence suggests that the state board of health advised them to drop the case.

[e]The precipitating issue here was the failure to report a contagious disease.

[f]A 1902 revision of the state statutes regarding the definition of the practice of medicine triggered this case.

and Mental Diseases Without [a] License" during the previous eighteen months. Mrs. Eddy herself wrote to Buswell with advice and encouragement. She referred him and his lawyers to John H. and Isabella M. Stewart, whose similar trial in Toronto, Canada, in 1889 had ended in their acquittal, adding, "Write me when you need me. Error has no power but to destroy itself. It *cannot harm you;* it cannot stop the eternal currents of Truth."[29]

During the subsequent trial the court heard about two days of testimony before the attorneys turned to their closing arguments. The prosecution emphasized the importance of protecting the public. E. N. Kauff-man, one of the attorneys for the state, stressed that "the defendant's religion was not on trial" but that the "imposition of their [Christian Scientists'] alleged curative practices upon innocent and helpless children, who were afflicted with scrofulous diseases or dangerous and contagious epidemics was a dangerous menace to the public health and safety."[30] Lead prosecutor Robert W. Sabin repeated the point: "This is a police regulation, not a religious persecution."[31] The attorneys for Buswell mounted a two-pronged defense. First, they argued that public safety could not be threatened because Christian Science healing proved safer than the methods used by the schools of medicine that sought to monopolize medical practice. Of the 136 death certificates filed in Beatrice during the previous eighteen months (excluding accidental deaths), supposedly only five had been treated by Christian Scientists and only two by Buswell. The remaining 131 had been treated by medical doctors. "Who," cynically asked defense attorney Alfred Hazlett, "constitute the state board of health which is to see that this law is enforced which is running this prosecution? They are four professors of medicine. The law should be entitled, 'a law to give a monopoly of the medical practice of the state to the allopaths, homeopaths and eclectic physicians.'"[32] Second, the defense asserted that Buswell had not practiced medicine. Turning to the jury, Frank N. Prout, a member of the defense team, spoke: "You are called to pass upon the right of this man to exercise his religious belief. In the face of the fact that he has cured by prayer a larger per cent. of his patients than any licensed physician in this community. The efficacy of prayer is as great today as it was in the time of Christ."[33]

The two local newspapers colorfully described the courtroom and its visitors, summarized in some detail the testimony of witnesses, condensed the arguments of the attorneys and the conclusions of the court, and offered glimpses of community attitudes.[34] Although the *Beatrice Daily Times* noted that "in some quarters public indignation has at times run high" during the last two years when Christian Science patients died, it declared that "it must be universally admitted that the members of the organization are numbered among our best citizens. They are peaceable, industrious and honorable, almost without exception."[35] Nonetheless, with the case in the jury's hands the *Daily Times* acknowledged that "there is a widespread prejudice against these people, but we are inclined to be lenient in criticism of a religious belief and practice that is founded upon the word of God. As to the legal status of the case the court will determine."[36] It is not known what Buswell called himself, but both papers referred to Buswell as "Rev." or "Pastor" and not "Dr.," which may have revealed the public's primary perception of his vocation.

Given such apparent public tolerance of the religious nature of Christian Science, the Gage County district court jury's decision to acquit Buswell may not have surprised the citizens of Beatrice. But the state excepted on the grounds that the trial judge had erred in instructing the jury that the intent of the legislation in question "was only to provide for the regulation of the practice of 'medicine, surgery, and obstetrics,' as these terms are generally understood." In the view of the state the legislature had not intended thus to restrict its meaning when it wrote, "Any person shall be regarded as practicing medicine within the meaning of this act who shall operate on, profess to heal, or prescribe for or otherwise treat any physical or mental ailment of another." Additionally, the state believed that since the court record made it clear that Buswell had received payment for his services, that confirmed that he had practiced medicine.

Although the state supreme court "conceded that the perfect toleration of religious sentiment, and enjoyment of liberty in all religious matters, is of paramount importance," it agreed with the district court and sustained its exceptions. Since the defendant had relied on the authority of the Bible for his defense, the court felt it not "amiss to refer to it for instances applicable to his case." It found the condemnation of simony in Acts 8 and the story of the transfer of Naaman's leprosy to Elisha's greedy servant Gehazi (2 Kings 5) instructive and concluded in light of them that "it is confidently believed that the exercise of the art of healing for compensation, whether exacted as a fee or expected as a gratuity, cannot be classed as an act of worship. Neither is it the performance of a religious duty, as was claimed in the district court." Moreover, the court concluded, "We find a very considerable part of defendant's brief devoted to an argument as to the inefficiency of the established and recognized modes of treatment in the cure of diseases, as compared with defendant's method, as tested by the results attained. The evidence upon which the case was tried convinces us that the defendant was engaged in treating physical ailments of others for compensation."[37]

Dismayed by the decision, Christian Scientists bemoaned the court's "harsh and unkind" treatment of Buswell and noted "the inability of the legal mind to interpret scripture." The *Journal* published an impassioned defense of Christian Science healing and a detailed critique of the decision in which it concluded that the court's conclusion represented nothing less than a "man-made" and "man-executed" effort completely to separate healing from religion and Christianity.[38] The medical community reacted very differently; the Nebraska State Medical Society received news of the decision with great pleasure.[39] James Vance Beghtol, a regular physician and member of the state's first board of health (established 1891), lauded prosecutor Sabin for his "plucky fight" and concluded that "he has done more for mankind in securing this decision than many whose services are rated at a high figure. So far as I am aware our Supreme Court is the first to hand down a decision on this subject, and it will surely establish a precedent to be followed in other states."[40] B. F. Crummer, chair of the society's committee on medical legislation, asked his colleagues "to pass a vote of thanks to the physicians of Gage County for their persistence in securing a ruling on this point, in the face of an adverse decision in their lower court. An eminent attorney in Omaha has assured me, after reading the syllabus of Judge Ryan's opinion, that this decision is bound to remain the law and guide in all such cases; and that simply means that every Christian Science healer may be fined repeatedly, as long as he or she insists on carrying on business."[41]

Although Beghtol believed that Nebraska's *State v. Buswell* (1894) had established a precedent that would be followed by other states, the decision in Rhode Island's *State v. Mylod* (1898) proved more influential. Under the direction of Gardner T. Swarts, secretary of Rhode Island's board of health, two policemen went to the offices of Christian Science practitioners Walter E. Mylod and David Anthony to seek treatment for feigned illnesses. After the visits, the state brought charges against the healers for the unlawful practice of medicine. The testimony revealed that they had not registered as physicians, had retained professional offices, and had called themselves "Doctor." Before the district court judge had reached a verdict, however, Mylod appealed to the state supreme court by challenging the authority of the state law requiring the licensing of physicians to curtail Christian Science practices on the grounds that it conflicted with the Constitution's protection of religious freedom.[42]

Rather than rule on the constitutional issue, the supreme court examined the intended meaning of the medical law in question to decide whether Mylod's practices constituted the practice of medicine. The unanimous court concluded that they did not, under the strict standard of the "ordinary acceptation and popular meaning" of the terms. In fact, continued the court, the Nebraska supreme court had found the opposite in Buswell only because the law in question had applied to any "treatment for physical or mental ailments" and not just the practice of medicine. "If the practice of Christian Science is the practice of medicine," concluded the Rhode Island court, then "Christian Science is a school or system of medicine, and is entitled to recognition by the state board of health to the same extent as other schools or systems of medicine. Under said chapter 165 it cannot be discriminated against, and its members are entitled to certificates to practice medicine, provided they possess the statutory qualifications."[43]

Although there was no chance that Rhode Island physicians would have tolerated such a solution—and the board of health immediately drafted a bill forbidding the practice of Christian Science—it would have suited the Christian Scientists just fine.[44] Rather than voicing their disappointment at the failure of the court to rule on the matter of religious freedom, the editors of the *Christian Science Journal* announced that if they were thus recognized as a school, then "it would be necessary that Christian Scientists should be represented on the State Board of Health, as only Scientists would be competent to pass upon the qualifications of Scientists. Such an arrangement would be quite satisfactory to Christian Scientists. We should find no fault whatever with such a 'medical regulation.' "[45]

Whether Scientists considered this a real possibility is unknown. They appeared equally satisfied with the Rhode Island court's assumption that Christian Science was a religion, something a court in Philadelphia had recently appeared to deny when it refused to grant a charter for a Christian Science church there.[46] Nonetheless, Eddy left no room for misunderstanding *her* current position that the Christian Science medical system was in direct competition with other medical systems. In an article published in the *New York Sun* on 16 December 1898, she granted that "in the ranks of M.D.'s are noble men and women, and I love them." But in a veiled threat she warned that "they must refrain from persecuting and misrepresenting a system of medicine, that, from personal experience I have proven to be more certain and curative in functional and organic diseases than their own—or we may not let theirs alone."[47]

As both the Buswell and Mylod cases showed, Christian Scientists proved themselves legally astute, willing to seek protection as practitioners of a system of medicine or as participants in a bona fide religion, whichever allowed them to continue their practice of religious healing. In fact, as in the Buswell case, they typically claimed both at the same time. Septimus J. Hanna, editor of the *Christian Science Journal* from 1892 to 1902, maintained that "Christian Science is a religion," but rather than simply defend its practice by appeals to a constitutionally protected exercise of religion, he argued that the new medical laws unfairly discriminated against nonorthodox healers and restricted the right of individuals to choose their own healers.[48] Hanna confidently asserted that "in a land of constitutional freedom, such as ours, there can be enforced no legislation that would take away the right of the citizen to select his own physician."[49] He believed that such a right was inherent and as sacred as the freedom to choose what one ate or wore, what one did with one's property, or where one went to school or church. This view that medical licensing invaded personal rights and created monopoly, a position also argued by Archibald McLellan, Hanna's successor at the *Journal*, persisted among Christian Scientists into the new century, but it grew more muted with the ascendancy of a First Amendment defense.[50]

Courts, rather than highlighting the question of religious freedom, rendered their decisions on the issue of medical practice. Sometimes, as in the Mylod decision, they ruled that Christian Science healing was not the practice of medicine. Practitioners Crecentia Arries and Emma Nichols of Milwaukee, Wisconsin, won acquittal on appeal in 1901 when they asserted that since they had administered no drugs, had performed no surgery, and had not even touched their patients, the law did not apply to their actions. The judge, agreeing, ruled that "prayer was not medicine."[51] On comparable grounds an Ohio court of common pleas acquitted Harriet Evans in 1898, and a Minnesota district court granted the demurrers filed on behalf of Mary Brookins and Albert P. Meyer in 1899.[52]

Less often, as in the Buswell decision, courts ruled that Christian Science healing was the practice of medicine. In *State v. Marble* (1905) the Ohio supreme court quoted extensively from the Christian Scientists' own writings to establish that they described their healing activities as "treatment"; thus, the court concluded, they practiced healing under the provisions of state law. Moreover, continued the court, although the state's bill of rights granted the right to worship God according to the dictates of one's conscience, it did not grant the right to practice religion in ways that harmed or endangered the public welfare. As precedent for this distinction between freedom of religious belief and religious practice, the court cited the Mormon polygyny case *Reynolds v. United States* (1878).[53]

Discussions of the issues involved in these cases abounded in the legal press.[54] After examining both the Buswell and the Mylod decisions, the editors of the widely circulated *Law Notes* summarized the dilemma of the courts: "Christian Science is, then, at once a religious belief and a system for the cure of diseases. It is this double aspect of the sect which involves the courts in difficulty."[55] The editors took the position that Christian Scientists had a constitutionally protected right to believe and practice their religion but argued that, since their religion was also a system for the cure of disease, they should not be allowed to invoke religious freedom to avoid obeying medical practice or public health laws.[56] William A. Purrington, New York City attorney and counsel to the New York State Medical Association, agreed that Christian Science was the practice of medicine but rejected the notion that medical practice laws forbade "the practice of Christian science, faith cure, voudoo, vitapathy, or any other 'pathy' or cult. Those laws provide only, at most, that no person shall practise medicine who has not pursued a course in medical study. There is nothing in them to prevent any licentiate from practising as he pleases."[57]

In Purrington's mind, such legislation did not establish a monopoly for a particular therapeutic school; it only determined the standards necessary to receive a state license. By implication, once that knowledge had been demonstrated, healers could practice as they liked. Following the precedents of the Mormon polygyny cases, Purrington carefully distinguished between the religious freedom to believe, which was constitutionally protected, and the religious freedom to act, which he believed was not constitutionally protected.[58]

Christian Scientists who participated in this legal debate also recognized the dual nature of their movement. However, they disputed the claim that they practiced medicine; for them the use of materia medica settled whether or not one practiced medicine. Carol Norton, C.S.D., member of the Christian Science Board of Lectureship, told the readers of the *Medico-Legal Journal* that "Christian Science is not the practice of medicine. It eschews the use of drugs. The courts invariably hold that medical practice must be understood as the physical use of drugs and medicines. Christian Science healers are therefore not practicing medicine, but carrying on the humanitarian work of alleviating human woe and healing the sick."[59]

As we have seen, Norton was wrong about what the courts had decided, and he did not tell the whole story when he claimed that Christian Science was not the practice of medicine. Only later in the article, while defending practitioners' charges for treatments, did Norton more fully explain that in his view they were "physician, minister, helper, and reformer" all rolled into one.[60]

Despite these and other encounters with the law, Christian Science leaders tried to reduce friction with the world as much as possible by keeping tight controls on their practitioners through regulations and supervision and by conceding certain legislative restrictions. Professional authority, as medical sociologist Paul Starr has pointed out, requires that professions achieve an "internal consensus" regarding the criteria for belonging and the rules and standards members must adhere to before it can carry "external legitimacy."[61] In the case of practitioners, this meant that Christian Scientists first needed to agree on principles of healing and teaching, a system of education that would perpetuate orthodoxy, and disciplinary procedures to ensure compliance. Then, having formed and preserved a professional self-image, they could present a united front as they took their healing ministry to Americans and sought to create a positive public image. In these matters they emulated their supposed adversaries, the physicians, who had worked to establish unified theories and consistent therapies, to upgrade medical education, and to regulate medical practice through licensing and examining boards.

During the late 1890s and early 1900s, Christian Scientists accelerated their efforts to coordinate and control practitioners and bring them into conformity with the evolving standards of professionalization in Progressive Era America. Christian Scientists claimed the right to establish their own licensing standards, since they possessed a distinctive medical system. Although Eddy had closed her Boston Massachusetts Metaphysical College in 1889, it had persisted as a paper institution, and she had continued to grant diplomas to her graduates under its name. The first edition of her authoritative *Church Manual* appeared in 1895, and in 1898 she created a Board of Education to standardize the training of practitioners. The church denounced specialization as "quackery"; the editors of the *Journal*, in order to exclude fraud, tightened their advertising policy for practitioners; and the Publication Committee more closely regulated the use of the titles C.S.B. and C.S.D.[62] At his trial, David Anthony, C.S.D., had unsuccessfully tried to convince the prosecutor that though his title stood for "Christian Science Doctor," it should be understood in the sense of a "Doctor of Divinity" rather than a medical doctor.[63] Not coincidentally, immediately after the Mylod decision (applied per curiam to Anthony), the 1899 edition of the Christian Science *Church Manual* unequivocally prohibited the use of the titles "Reverend, and Doctor," unless received "under the *laws* of the *State.*"[64]

No matter how much Christian Scientists lobbied and published defenses, they had difficulty defusing the negative image and even ill will often surrounding the fact that Christian Science practitioners charged for their treatments. Therefore, in an effort to "raise the vocation of Scientists from being looked on by the world as primarily a means to a livelihood," Christian Scientists opened missions and dispensaries that provided free treatments and literature to the worthy poor.[65] Patients with acute problems received immediate attention on the premises, and dispensary workers referred more serious cases to practitioners in the community. Twenty-nine primarily urban dispensaries were established across the country within one year of the decision at the Association's 1889 annual meeting to encourage such service. However, like the medical community's urban dispensaries, which, according to Charles Rosenberg, "lost much of [their] appeal for the medical elite," by the turn of the century tensions between dispensary personnel and community practitioners led to the demise of Christian Science dispensaries as treatment centers.[66] For many practitioners Science *was* a "means to a livelihood," and although dispensary practitioners did begin to charge patients based on their ability to pay, their prices still squeezed the profits of those in private practice. To avoid further strife, many dispensaries halted treatments and became Reading Rooms, which distributed literature, created a temporary retreat from hectic urban life, and provided a place "not only [of] peace and quiet, but healing as well."[67]

Christian Scientists and Physicians Lobbying State Legislatures

As the courts struggled to determine the legal status of Christian Science practice, during the 1890s both the medical community and Scientists mounted far-reaching campaigns to persuade the public and its legislators to support their respective views. Physicians lobbied legislators to explicitly include Christian Scientists and other religious healers in medical practice acts, and Christian Scientists worked to exempt themselves from such legislation.

The medical community took the lobbying machine it had already used to pass much legislation and turned it against Christian Scientists.[68] The committees on medical legislation of the state medical societies tried to convince legislators that anyone who practiced medicine should conform to certain standards that would ensure the safety of the public. Christian Science healing, they argued, constituted the practice of medicine; therefore its practitioners and all religious healers who received payment for their treatments or "prayers" should be required to demonstrate at least a knowledge of anatomy, physiology, and medical diagnosis. *JAMA* even encouraged its members to follow the example of Michigan and have "candidates for the legislature . . . called on before election to state their positions as to the legal toleration or recognition of these [Christian Science] practices."[69]

However, the medical community had to overcome a mixed public image before its efforts could bear much fruit. During the not-so-distant past the physicians had been engaged in sectarian struggles for supremacy, and it often appeared to the public that medical legislative efforts sought monopoly more than public protection. Also, unless it worked carefully, the medical community would appear to be trampling on the religious freedom of Scientists for the selfish purpose of keeping patients for themselves. The public could turn Scientists into martyrs, with the unintended result that religious healers could gain more freedom to do as they pleased. These realities, coupled with the nonmaterial nature of most religious healing, led a few physicians, such as Chicago's A. S. Burdick, to doubt that the struggle was even worthwhile. He thought that "to require examinations in anatomy, pathology, etc., is therefore absurd. Non-interference with their practices, so long as it is a matter of individual choice and does not violate other laws or jeopardize other persons[,] is advised."[70]

To remedy their poor public image, some physicians encouraged the profession to break out of its elitist mold and work with the public press to educate Americans about the scientific basis of health in a modern world. In a forceful call to arms, Milwaukee physician Ralph Elmergreen pled:

We must marshal our latent forces and strike. We must assert the rights of a state. With the bayonets of intelligence fixed to trammel down prejudice, and the sights of reason focused to kill ignorance, we must press forth and make the secular and religious press our battle-grounds, the school-house, aye, the church, our reinforcement.

Our future lies in education. The masses must be educated in medical science. . . . We have no popular medical literature, no pamphlets popularizing science; no leaflets of convincing medical statistics setting forth the value of certain prophylactic or sanitary measures; no propagandic newspaper-backing to answer the brazen effrontery of the medical mountebank—and in this lies the explanation of our failure to win the popular ear.[71]

Some evidence indicates that such efforts to popularize the science of health improved the public's understanding and trust of medicine, but their effect on attitudes toward Christian Science remains unknown.[72]

While physicians pulled on one sleeve of the legislators, Christian Scientists tugged at the other. Scientists recognized that "the standpoint from which Christian Science has been most vigorously and persistently assailed is that which denies its efficacy as a system of healing"; thus they lobbied legislatures to have religious healers, or more explicitly Christian Scientists, exempted from medical licensing laws.[73] To coordinate this effort, which began in earnest during the late 1890s, Eddy founded the Committee on Publication in 1898 to correct errors in the public press and to lobby government for legislative relief. First comprising three members, in 1900 the committee came under the management of a single person responsible for overseeing the activities of all the Committees on Publication (all one-person, male committees) assigned to each state and several large cities. Alfred Farlow, C.S.D., served as manager from 1900 to 1914, followed by Judge Clifford P. Smith, C.S.B., from 1914 through 1929.[74]

The COPs, as they became known, defended the healing activities of practitioners, explained their beliefs and the Christian roots of their practices, and cited evidence of their wonderful cures and the failures of medical doctors. They also spearheaded a critique of physicians for the way they practiced medicine and for their monopolistic efforts that grew so contentious that Eddy had to rein them in, even though her remarks in the *New York Sun* just a year before had probably egged them on. She penned a new 1899 church bylaw that warned her members not to "publish, nor cause to be published, an article that is uncharitable or impertinent towards religion, medicine, the Courts, or the laws of our land,—on penalty of being removed from the Editorial corps, and the Board of Lectureship."[75] Although even the editors of *JAMA* admired the effectiveness of the COPs, they whined that "if the medical profession maintained a publicity department that cost a hundredth part of the 'Christian Science' press agency, hands would be raised in holy horror and from the house-tops would come the cry: The very foundations of our civil liberties are threatened."[76]

The struggles in Missouri over new 1901 legislation to regulate the practice of medicine and surgery illustrate the way physicians and Christian Scientists battled into the twentieth century.[77] Disgusted by actions taken by the governor against regular physicians, members of the Missouri State Medical Association resolved at their 1897 annual meeting "to enter politics," predicting that they "would put in the gubernatorial chair, a man who would respect the voice of the 6000 physicians of our State."[78] This they accomplished in 1901 with the election of one of their own, Dr. Alexander M. Dockery, to the governor's office in Jefferson City. But it did not solve all their problems, for with the presence of Weltmerists (a variety of magnetic healers), osteopaths, and Christian Scientists, Missouri had, in the view of one observer, "more than its share of these fantastic schools."[79] In an effort to corral these "isms and fads," the association mounted a campaign in 1900 and 1901 to secure the passage of the "Hall medical bill." The bill proposed the regulation of not only those who "practice medicine or surgery" but also those who "profess to cure and attempt to treat the sick and others afflicted with bodily or mental infirmities," which would have included religious healers and Christian Scientists. Moreover, to prove their qualifications as practitioners, applicants would be required to pass an examination in "anatomy, chemistry, physiology, pathology, therapeutics, obstetrics, gyne[c]ology, surgery, practice of medicine, medical jurisprudence and hygiene and such other branches as the state board may direct," subjects no self-respecting Scientist would deign to study.[80]

Working with the association's legislative committee, U. S. Wright, president of the association, reported making many calls on prominent men over the State, by correspondence, and telegram, asking them to be present in Jefferson City, at important dates in the progress of the medical bill, before the House and Senate, and had letters written to important characters in the Legislature, hoping to give them more light, and twice left the bedside of my dying mother, spending several hours each day at the Capitol, with the Committee, in their efforts to induce the Representatives of the House and Senate, to look favorably on our cause.[81]

In addition, physicians mounted a statewide petition drive that targeted the members of the Presbyterian, Baptist, and Methodist denominations, apparently in an effort to rally their support for the bill and thereby avoid the charge that only physicians objected to religious healing.[82] Others spoke for the bill, including Rev. Dr. W. B. Palmore, editor of the Methodist *Christian Evangelist* of St. Louis, who "did not believe Christian Scientists were as full of voodooism as the 'magnetic healers'" but felt they were "bad enough, and as this bill would catch the magnetic healers, it ought to be passed."[83]

Christian Scientists responded with their own lobbying campaign that culminated with their descent on Jefferson City by the trainload to confront the jointly assembled committees on public health of the state assembly. Estimated to number about one thousand, the Scientists sat on the floor and stood in the aisles from 7:00 P.M. until midnight on 5 February 1901 to await their chance to oppose the legislation. Among those who addressed the committees, Frank Cooper, a Scientist from Kansas City, reminded the legislators of the Constitution's protection of religious freedom and testified that if the bill passed it "would wipe out every Christian Science church in Missouri." However, the physicians on the committees, including an osteopath, "went after Cooper vigorously, particularly when he admitted that the Christian Science healers sometimes charge for services" that committee members felt should be a "free gift of God." Trying a different but familiar tack, Scientist A. H. Dickey, also from Kansas City, "urged that the Christian Scientists would be satisfied if the legislature would give them representation on the board of health." This, of course, would have allowed a Christian Scientist to examine Scientist applicants and thereby moderate the effects of the other medical sects' antagonism.

When the bill came up for final consideration in the assembly, Representative James A. McLane and Senator William H. Haynes offered in their respective houses several amendments that would have exempted Christian Scientists from the new regulations, but all failed. The House approved the final bill seventy-seven to forty-four, as did the Senate, twenty-six to five.[84] Although political party allegiances apparently played no significant part, reports made it clear that both sides had appealed to religion. Not only had Christian Scientists complained about religious persecution, but assembly representatives with membership in the Presbyterian, Baptist, or Methodist churches reported that "their ministers urged them to support the bill." Roman Catholics, fearing that administering extreme unction, which occasionally resulted in healings, would be illegal if the bill passed, unanimously voted against it.[85]

Similar struggles occurred in states throughout the country, with results that varied from place to place and time to time. One scholar has referred to 1900–1915 as the period of "legal recognition" for Christian Scientists.[86] In a flurry during the few years before 1905, thirty-eight states saw legislation introduced that either prohibited Christian Science practice or forced practitioners to comply with medical practice acts.[87] Legislators passed such bills in Oklahoma (1899), Colorado (1903), and Nebraska (1905), but the governors vetoed it.[88] Among the states whose legislatures passed and governors signed such legislation were Virginia (1903), Maryland (1908), and Ohio (1915).[89] Increasingly, however, governors and legislators explicitly exempted Christian Scientists from compliance with their medical practice acts, as was the case first in Illinois (1898) and then in numerous other states, totaling twenty-eight by 1917.[90] Regarding the narrow issue of the practice of medicine, the trend clearly was toward exempting religious healers. Nonetheless, the matters of public health ordinances and child welfare remained very much up in the air.

At times these legislative attempts to clarify the legal status of religious healers only further confused the courts. In 1907 New York broadly defined the practice of medicine to include anyone who claimed to be able to "diagnose, treat, operate or prescribe for any human disease, pain, injury, deformity or physical condition." However, the law also stated: "This article shall not be construed to affect . . . the practice of the religious tenets of any church."[91] While the legislators had intended to define the practice of medicine broadly enough to include restrictions on marginal healers, they also wanted to explicitly exempt religious healers of recognized denominations.

Physicians continued to chafe at the exclusion of "offbeat" religious healers, notably Christian Scientists, because in their view such practitioners threatened health and safety and defrauded the public. Algernon Thomas Bristow, editor of the *New York State Journal of Medicine*, believed that "religious liberty does not mean religious license" and that "the practitioners of Christian Science should be compelled to pass the same examination that other practitioners pass willingly, and they should not be allowed to evade the law and amass fortunes from their dupes under the sneaking and hypocritical plea that mental healing is a religious act and that, therefore, they should not be interfered with in their insane and stupid mummeries."[92]

Wanting to test such beliefs, the New York County Medical Society filed a complaint in 1911 against Willis Vernon Cole, a New York City Christian Science practitioner, for the unlicensed practice of medicine. In an opinion greatly dependent on the brief filed by Almuth C. Vandiver, counsel for the county medical society, Justice J. J. Freschi bound Cole over for trial on the grounds that although he had the "right to believe that he can heal by prayer," he must subordinate that right "when the free exercise of such belief either impairs and endangers the health of the people or tends to place their health in jeopardy."[93] In so writing, the judge sought to distinguish between a protected freedom of religious belief and a restricted right to practice religion. Cole's first trial ended with a hung jury in 1911, but on retrial in 1912 a second jury found him guilty, launching an extended appeal process.

A divided appellate division court affirmed the jury's decision in 1914, but in 1916 the state court of appeals overturned that ruling. The unanimous court ruled that the statute's exception for religion "is not confined to worship or belief, but includes the practice of religious tenets" when "practiced in good faith" and not simply "for the purpose of thereby maintaining a business and securing a livelihood."[94] Christian Scientists greeted the decision with a feeling of vindication, but they had not stood idle while the courts determined their fate. At their importuning, in 1914 the New York legislators had passed a law that explicitly exempted Christian Scientists from compliance with the medical practice act. Under the threat of a veto by the governor, about two thousand Scientists had arrived in Albany to convince him otherwise, but to no avail. Governor Martin H. Glynn had concluded that if the bill

> simply allowed Christian Science healers to practice among the followers of Christian Science, there might not be serious objection to it, but it goes further than this and adopts the loose language of the Medical Practice Act of Illinois, where the standards of state education and public health are distinctly lower than they are in the State of New York, and it opens the gates to all kinds of medical pretenders who, as a matter of fact, "administer who, as treat the sick or suffering without the use of any drug or material remedy," and who, if this bill were approved, would swarm across our borders and pretend to practice medicine upon our citizens. Under the phrasing of the proposed law, I am precluded from passing upon the claim of the sincere believers in Christian Science, and I am, therefore, constrained to disapprove the bill.[95]

In 1908 Alfred Farlow, manager of all COPs, summarized well the dilemma for Christian Scientists that persisted into the twentieth century. In his response to legislative pressures to control the activities of practitioners, he declared that "healing the sick is a consequence of Christian Science practice and not its prime object. The practice of Christian Science is not a business, but a ministry, not a profession, but a rule of life."[96] Farlow may have been partly right; at least he was correct in seeing that Christian Scientists increasingly wanted to be seen by the larger American society as reform-minded, freedom-loving Christians whose healers, safe within their First Amendment rights, worked only as ministers and not as physicians. But this self-chosen public image, which American courts increasingly accepted as the truth, reflected a changing emphasis in the movement that differed from the realities of its historical roots. Disdaining much of their sectarian heritage in both religion and medicine, Christian Scientists sailed into their future with a rewritten past based on an image of apostolic purity and purged of medical sectarianism.

But Farlow was also partly wrong. Although Christian Scientists may not have wanted the public to think of them primarily in terms of their business activities, those Scientists who earned a living as full-time practitioners thought seriously about the business and ethical aspects of their labors and sought to improve their status by controlling and regulating their profession. Despite the disclaimer above, Farlow, a practitioner himself, spent much of the space in his article defending the rights of Scientists to heal the sick and to earn a living from it. "Healing the sick" theoretically may not have been the "prime object" of Christian Science, but it undoubtedly still drew the greatest number of believers into the church, provoked the most intense opposition by the medical community, and presented the clearest distinguishing feature of the movement to the public. Although Christian Scientists might have eased such public pressures by deemphasizing physical healings, many feared that thereby the church would lose much of its appeal. In the opinion of influential lecturer Carol Norton, Christian Science prospered "most through the genuine results obtained in regeneration and physical healing."[97] But slowly the focus of the public debate over Christian Science healing moved from concerns about monopoly, "fairness," and therapeutic choice to anxieties about public health and the protection of children. Even then, the struggle remained more than simply a matter of conscience.

7

Public Health and the Protection of Children

Christian Science, however, has much to answer for in the fatal cases of contagious diseases, and the deaths among its devotees from lack of nursing and hygienic measures, all of which are considered unnecessary. While the adult, being a free moral agent, may rely on faith alone to resist death, it is a little less than sacrilege to allow innocent children to die for the lack of medical attention and good nursing.
       —L. Watkins, M.D., 1903

It must not be forgotten that it has never been proved, and I hold it incapable of proof, that Christian Science treatment is inadequate and accordingly to the child's detriment.
    —Albert F. Gilmore, C.S.B.?, 1920

Physicians persistently denied the charge that they sought to criminalize religious healing as part of their scheme to establish a medical monopoly. They protested that they had nothing against religion and had only the health and well-being of society at heart. In 1899 many if not most members of the medical community may have agreed with the editors of *JAMA* that "'Christian Science,' Dowieism, and such fads" would be "comparatively harmless delusions" if they affected only their responsible, adult adherents. "But," the editors continued, "they are more than this, they are Molochs to infants, and pestilential perils to communities in spreading contagious disease. It is therefore the duty of the medical profession to do what it can to enlighten the public in regard to them."[1] Christian Scientists, however, disputed the claim that scientific medicine held the key to improving the public's health and protecting children; some of the general public shared their doubts.

After the Civil War, Americans witnessed a great influx of immigrants and a resultant population explosion that contributed to the transformation of America from a primarily rural, agrarian society into an urban, industrial power ready on the eve of the Great War to take its place on the world stage. Many Americans prospered, but countless others lived in its increasingly impersonal and dangerous urban landscape. In the words of one historian, a close examination of America's cities exposed a fetid cauldron of "infectious diseases; crowded, dark, unventilated housing; streets mired in horse manure and littered with refuse; inadequate water supplies; unemptied privy vaults; open sewers; and incredible stench."[2] Much of the history of public health in Progressive Era America is the story of how Americans worked to clean up that mess.

Sanitation proved to be one of the most obvious public health initiatives because of the ever present piles of malodorous and unsightly garbage, manure, and dirt and the popular notion that filth spread disease.[3] Such sentiments about the connection between filth and disease also coincided neatly with a growing zeal for personal hygiene as a means of preventing disease.[4] Not surprisingly, therefore, one of the early matters on which Christian Scientists appeared to offend a growing public health consensus was sanitation and personal hygiene. Given their metaphysical rejection of the body and its physical surroundings, some observers wondered why Christian Scientists would need to eat, let alone pick up their garbage, and many doubted they practiced personal hygiene. In the words of one critic, "The school [of Christian Science] rejects all sanitary science. Hygiene is denounced as a most culpable concession to evil. All physical protective measures, all correction of putrescence, all shutting out of poisonous matters, all fighting of filth, is wrong."[5] Scientists vigorously disavowed such distortions, asserting that "in so far as hygiene and sanitation denote purity and cleanliness, Christian Science is in full accord."[6] But contagion proved a different matter altogether.[7]

Public Health and Contagious Disease

On 3 June 1895 police officer James McManus served Christian Scientist Charles A. Owen with a warrant for his arrest and brought him before J. A. Le Claire, police magistrate for the city of Davenport, Iowa. Le Claire charged Owen with "violating Section 7 Chapter 33 of the Ordinances" of

Davenport by neglecting "to report to the Board of Health or Police Station a case of Dip[h]theria which he was called on to attend within the bounds of the City within the time prescribed."[8] Three days earlier at 10:30 in the morning six-year-old Alma Bohnhof had died of diphtheria at the home of her parents in the city's west end. City health officers had already been concerned with the high incidence of contagious diseases in that region of the city and were anxious to keep a close count on new cases; but they were not just worried that Alma's case had not been reported to the proper officials. Of at least equal concern was that she had been treated by Charles Owen, pastor of the First Church of Christ (Scientist) in Davenport and a Christian Science practitioner in the area for the past two or three years. According to the *Davenport Sunday Democrat,* Owen had already been in trouble with the law when his sister-in-law and niece died recently of diphtheria while under his care. The "Christian Science 'doctor,'" the *Democrat* concluded, "seems no more able to cope with this deadly malady now than formerly. In fact, he is unable to recognize it at sight or after long acquaintance."[9]

This seemingly blatant flouting of city health ordinances provided city physician A. W. Cantwell with an opportunity not only to warn citizens of the dangers of Christian Science treatment but to publicize health problems in the three southwestern wards of the city. At the 5 June meeting of the city's board of health, Cantwell reported that the second half of May had seen ten new cases of diphtheria from the three lower wards and that nineteen homes had been placarded for membranous croup, scarlet fever, or diphtheria. Although unsure why there was a higher incidence of disease in these wards, Cantwell advised the city to connect the region's schoolhouse to a sewer and eliminate common drinking cups at schools. P. A. Radenhausen, a physician resident in the area, encouraged the city to supply running water to the schools so that "by the flow of water from a hydrant the cups would be kept clean."[10] It was within this climate of public opinion that Justice Le Claire set Owen's trial for 10:00 A.M. on 6 June 1895.

Charles Owen's trial in the police court of Davenport, Iowa, resulted in a judgment of "guilty as charged" and a fine of $50 plus court costs. The *Democrat* expressed surprise that Owen had "quite a number of followers here who are not only thoroughly in sympathy with the [Christian Scientist] ideas, but who are actual sharers of his hallucinations," and it marveled at the tolerance of Davenport when most places would have proclaimed Christian Science healing "a fit subject for the beneficent operations of the foolkiller, if not of the insanity commission."[11] On appeal the district court overturned the conviction on the grounds that the city failed "to prove that Owen knew diphtheria from sore throat," but in the meantime the city board of health had made good use of the case's notoriety and had gone far in establishing itself in the mind of the public as the "guardian" of Davenport's health.[12]

To fight infectious disease, postbellum states and many cities, like Davenport, had passed public health ordinances to standardize and refine the collection of the vital statistics necessary to track the origin and spread of contagious disease and to authorize quarantines when necessary.[13] For example, Massachusetts enacted legislation in 1884 that required the family and attending physician of a patient infected with "small-pox, diphtheria, scarlet fever or *any other disease dangerous to the public health*" to report it to the selectmen or the local board of health. Then in 1893, reflecting a nationwide trend to expand the authority of state boards of health, the state "required local boards of health to notify the State Board of all dangerous diseases at the time of diagnosis." By 1907 the state board, which had assumed total authority to determine which diseases posed a threat to public health, required the reporting of sixteen dangerous diseases.[14]

It soon became apparent to most states, however, that it would be difficult if not impossible for laypersons such as family members to diagnose dangerous diseases accurately; hence the authorities focused on ensuring that physicians and other healers complied with the requirements. But even here officials met some resistance from those who doubted the theory of contagion or distinguished between "infectious" and "contagious" diseases and from those who believed that collecting such statistics only wasted their time and had no effect on their choice of proper treatment. Before such requirements could prove effective, therefore, there needed to exist the expertise necessary to make an accurate diagnosis and a widespread belief in a comprehensive doctrine of contagion.[15] The medical community and public health officials grew to believe that medical licensing would be the best way to ensure the necessary expertise. Chicago physician A. S. Burdick spoke for many in 1901 when he wrote: "The desire to conserve the public health undoubtedly stands at the head among the reasons for desiring state regulation of medical practice, although it is doubtless true that some medical men think more of the restriction of competition." He granted that Christian Scientists and faith healers should be allowed to practice unhindered, "so long as it is a matter of individual choice and does not violate other laws or jeopardize other persons"; nonetheless, he concluded that they should "be compelled to obey the laws which require the reporting of births and deaths, the notification and quarantine of contagious diseases, etc."[16]

When even physicians acknowledged the mixed motives behind public health legislation and supported therapeutic liberty, it is not surprising that Christian Scientists tried to make public health regulations an issue of individual rights. Septimus J. Hanna, then editor of the *Christian Science Journal*, testified before a Massachusetts Senate committee that the proposed 1893 public health "bill is so palpably in violation of personal rights and personal liberty that even under the police power of the State it could not be enforced."[17] In Hanna's view physicians unjustly raised public fears about health dangers in order to get legislation passed that would establish a medical monopoly for them. Such views persisted well into the new century, but as time passed and the public grew more fearful of infectious disease, fewer Americans seemed to have taken such arguments seriously.

While Scientists rallied for personal liberty, members of the medical community labored to educate the public and unpersuaded physicians about the danger of germs and the wisdom of public health ordinances regarding infectious diseases. Richard Olding Beard, M.D., from Minneapolis encouraged the Minnesota medical community in 1903 to "voice in local and state conventions the consensus of medical opinion in respect to the subjects [of public health] and the form of proposed [legislative] measures. It must carry on a campaign of education, firstly, of its own members; secondly, of the public at large; and, lastly, of its legislative representatives."[18] In these efforts to convince the public of the medical community's competence to determine public health dangers, physicians often laid claim to scientific authority over matters of disease and health. Capitalizing on the public's supposedly growing confidence in science, physicians argued that science had lifted medicine above the arcane interschool debates of yesteryear and had placed it on a bedrock foundation of truth.

Christian Scientists, who seriously doubted that physicians had finally discarded their disputatious ways, preferred to believe that, as one author put it, "the fickleness of Materia medical error, shown in the celerity with which it discards its working basis of yesterday to adopt a new one tomorrow, cannot fail to raise, among thinkers, the question as to whether *Materia Medica* has any real basis at all."[19] "On the other hand," he continued, "Christian Science pursues the even tenor of its way; working yesterday, today, always, upon the same changeless basis, *viz:* Immortal Mind."[20] Scientists also charged that medicine was an inexact science, despite its recent claims. Hanna asserted:

> If it is worthy this distinction, it can show that, as a system, it can cure all manner of sickness; that it loses no cases; that not a death occurs under its ministration except by accident or as the result of senility; that, in its hands, infancy and youth are exempt from death; that if these ends are not accomplished, it is because of incompetent disciples or circumstances which make a fair exercise of its powers impossible, and not inherent fault or lack in the system or its science.[21]

In contrast to the apparently willy-nilly empiricism of medicine, Christian Scientists held that their system was based on immutable and unchangeable divine law. To hammer home their point, Christian Scientists took pleasure in noting that a "relatively large number" of physicians had joined their ranks to offer "among the most valuable testimonies we have to present."[22] Clearly Scientists recognized the propaganda value of a turncoat in time of war.

Only superficially, however, did it appear that Christian Scientists and medical scientists primarily argued with each other about the cure rates and conversion records for their respective systems. More fundamentally, they argued past each other about the relative truth value of Science versus science. The core of the Christian Scientists' argument for their healing system ultimately rested on a metaphysical first principle— God is Good—which no amount of empirical evidence could refute. And medical scientists praised their own new first principle—germ theory— which needed to be continually tested by empirical evidence. Given these competing first principles and contradictory epistemologies, there remained little ground for constructive dialogue.

But in the public arena, where metaphysics and epistemology usually took a backseat to common sense, the nature of the debate changed. As the two groups vied for the support of the American people, they couched their arguments in terms that meant something to ordinary Americans, and phrases such as liberty, cure rates, and public health danger assumed a crucial role in the struggle for ascendancy. This public struggle between Christian Scientists and physicians over the definition of "true" healing bore many similarities to the political struggle two decades later between fundamentalists (creationists) and scientists (evolutionists) over the definition of science. Both Christian Scientists and fundamentalists, like many American populists, sought to establish authenticity for their ideas by a "vote," not by appeal to the "private" standards of a self-appointed elite.[23] However, although Scientists' populist appeal to individual liberty and their critique of medical inconsistencies worked to their advantage for a while, their contrarian views of the material world ultimately proved their undoing. Physicians held a distinct advantage because very few Americans doubted realist assumptions about the existence of a material world. Germs were an elusive entity for many, but in the final analysis the nonexistence of matter proved less believable.

Beyond the often lamented affront to personal liberty and the argument over science lay legislative terrain much more threatening to the integrity of Christian Scientists. The requirements that medical practitioners diagnose and report contagious diseases, report the cause of death when a patient died, and sign death certificates placed them in a catch-22. Because they did not diagnose diseases in a way that would allow them to turn in acceptable reports, the requirements appeared to force them either to disobey the law or to quit practicing Christian Science. For Davenport's Owen the problem was not so much that he had been "unable to recognize" diphtheria, as that he had refused to recognize it because to do so would have ensured its presence.

In 1897 Kansas City health officials similarly charged practitioner Amanda J. Baird, a pioneer Scientist in that city, with failing to report "a most malignant type of diphtheria" that had resulted in the death of a ten-year-old girl. Speaking for his fellow physicians, city health officer Dr. Ernest von Quast noted that "Christian Scientists have violated the sanitary laws and the city ordinances constantly, and for a long time." But this time, he declared, "I shall see that every one, including Mrs. A. J. Baird, is prosecuted to the full extent of the law."[24]

The threat, however, did not seem to shake Baird's confidence that she was in the right. When a reporter asked about her failure to report the disease to the health office, she responded: "We Christian Scientists are living according to our own spiritual laws, as I said before, and when we conflict with human laws we disregard the latter. We don't believe that the health department ought to meddle and put up cards. When a card bearing the word diphtheria is displayed on a house, it frightens the people that see it and does an infinite amount of harm. But what harm can it do if the card is left off the house?"[25] Clearly, Baird recognized the generally held belief in and fear of contagion, but she refused to alter her actions to accommodate the public. Instead, she took the opportunity to evangelize for her beliefs and to chide the community about the practice of posting and quarantining, which in her view only exacerbated the dangers to public health by inciting fear.

When news of what had happened in their neighborhood reached the residents, it "caused a small panic." As the *Kansas City Star* reported, many homes surrounded the infected household, "most of them containing children whose mothers are intensely anxious for the safety of their little ones and angry not only at the flagrant violations by Christian Scientists of the law the city makes for the protection of the public health, but of the ignoring of all respect for other people's opinions." As one neighbor put it, "If they wish to treat their children according to their own methods—that is, not treat them at all—very well; that is not the public's business. But the moment they threaten the public's health it certainly does become the public's business to protect itself."[26] Fearing that their sick children had diphtheria, anxious parents swamped doctors' offices, and the board of education acted quickly to allay further alarm by ordering the fumigation of area schools.

The *Star* took a firm editorial stand against "the peculiar notions of the people who regard the [public health] ordinances an infringement upon their religious liberty." The paper argued:

No religious idea or belief can or will be countenanced in this country which conflicts with the authority of the state or sets out to defy its laws. Our system of government, in all of its details, allows the largest measure of liberty to the citizen which is consistent with the safety of his neighbor, but beyond that it cannot go. While the right of Christian Scientists to withhold medical treatment from members of their own faith may be open to debate, there can be no reasonable difference of opinion as to their responsibility to those regulations which are provided in wisdom for the preservation of the public health and the restriction of pestilence.

The ordinance requiring the marking of houses in which there are contagious disorders is founded upon the rational assumption that the spread of diphtheria and scarlet fever and the like cannot be prevented by calling these diseases whims and notions, and it should be vigorously enforced.[27]

As far as the *Star* was concerned, all Christian Scientists believed and practiced the same, but this was far from the truth. Discord and dissent had dogged the movement's every step in Kansas City since its establishment there in 1886 by Emma D. Behan. After the First Church of Christ, Scientist, had blown up over doctrinal and institutional debates that mirrored tumults at the Boston church, Mrs. Baird had founded in 1890 what became the city's Second Church. Alfred Farlow, C.S.D., a graduate of Eddy's Normal class, had arrived from Topeka, Kansas, in 1892 to set up practice with his brother and sister, and together they had started their own services that grew into the Third Church in 1895. Baird's arrest occurred just as the unified First and Second Churches and the Third Church had nominally resolved new disputes over their competing building projects, but organizational and doctrinal differences and personal jealousies still festered.[28]

These church family squabbles spilled into the public after Baird's arrest when some Scientists publicly objected to her rather cavalier attitude toward the city's sanitary laws. Farlow told a reporter that he did not agree with the radical position Baird had taken and that he would not have permitted the dead little girl's brother "to attend school while his sister was sick with this malignant disease, that was very wrong." Although quick to say that it was only his own opinion, he added: "I don't believe the doctrines of Christian Science will bear her out in what she has done. She should not have been so—so high and mighty as she appears to have been. We Scientists are not yet so strong that we can defy all human laws. As for myself, I would hesitate a long time before I would refuse to notify the board of health in the case of a malignant disease I was positive about."[29] William B. Dickson, another Christian Science practitioner and teacher, concurred. He believed that "the law in regard to notifying the health department in case of malignant disease is not an outrageous law." Although he could not be sure, he thought that in this case he "would have notified the health department."[30]

Farlow had forged this position through tribulations of his own. Earlier in his career a child had died while under his treatment. Afterward the authorities had urged the child's father to turn Farlow in if he treated another case of infectious disease, but Farlow had persuaded the father to break that promise after a second child died because Farlow had feared "cruel treatment and persecution from the hands of physicians" if the affair became known. Pricked by guilt, Farlow subsequently visited with the town's health officer and coroner; having found them understanding, he regretted having influenced the father to break his word. He concluded that this "experience has taught me many useful lessons and has given me more charity for honest physicians who are conscientiously trying to fulfill the duties entrusted to them."

But Farlow had not lost confidence in Science. He still believed that "Christian Science is not only a good remedy for the patient, but is the very best known disinfectant. All the known means for preventing the spreading of disease taken together will not compare with the understanding of a Christian Scientist." And he undoubtedly agreed with Carol Norton that Science should be "judged by its uniform successes, not by its isolated failures."[31] But he had gained a new respect for the many well-meaning physicians who worked for ends similar to his own, albeit with different beliefs, and it dawned on him that they and the general public had rights too. "The time is at hand," he concluded, "when some of the laws of our land need to be changed to suit more scientific methods [Christian Science]," but we must be "careful to recognize the rights of others, and not seek too rapidly to enforce new ways before their time."

Through the deaths of his patients he had also gained a new appreciation of the limits of his practice. As he stated, "Christian Scientists demonstrate absolute Christian Science as far as possible. Beyond this we are obliged to choose the lesser of two evils." But the mounting number of Christian Scientists on trial made him sense that the public not only noticed Scientist failures but had grown increasingly intolerant of them, and he needed to mollify the community. Evidently Eddy approved of this more moderate approach to the world, because she tapped Farlow to join the three-member Committee on Publication in 1898 and then made him its single-member manager in 1900.[32]

In contrast, Baird took an older, more defiant stand for truth. Unrepentant, she denied that she or anyone else associated with the treatments had even thought about diphtheria, let alone diagnosed it, and she asserted that "it would be inconsistent with our belief to place a sign of diphtheria on the house when we believe that there is no such thing as diphtheria." In a thinly veiled reference to Farlow, she acknowledged that "I am made out a breaker of the laws of the land, but that's a mistake. . . . We do not feel bound to the laws of hygiene, but to the laws of God."[33] Like Owen's, Baird's conviction was overturned, although it took an appeal to the state's Kansas City court of appeals to reverse the decision in 1902 on the familiar grounds that the city had failed "to prove that a *physician* attended the sick person and that he *knew* the case was one of those diseases mentioned in the ordinance."[34]

Not only in Iowa and Missouri but across the nation, from Maryland and Rhode Island to New York and Minnesota, Americans learned of the Christian Scientists' peculiar beliefs about infection and of their failure to report contagious diseases.[35] They also learned of the ways practitioners tried to comply with or circumvent the requirements involving death certificates. Some Scientists raised more than a few eyebrows when, assuming the role of physicians, they signed death certificates and indicated that the cause had been "sin and fear," their common explanation for death.[36] Other practitioners, recognizing that if they signed the certificates they might jeopardize their claim that they were not doctors, persuaded accommodating physicians to sign them. Some of these cooperating physicians had converted to Christian Science, but not all.[37] If the complaints by physicians in their medical journals are any indication, this cozy relationship between practitioners and some physicians began about 1899 and continued into the 1920s.[38]

Despite the public protests by Christian Scientists against the obligation to report and quarantine infectious diseases, the arrest and trial of Owen was apparently the first of only three cases before 1920 in which Christian Scientists were tried for breaking such a public health ordinance. Usually, rather than forcing practitioners to obey infectious disease ordinances, public health authorities tried to keep them from treating any diseases at all. Therefore, when a case before the turn of the century involved an infectious disease, officials customarily prosecuted on the grounds of unlicensed practice (see appendix). In such cases many of the familiar arguments against medical licensing reappeared, including the supposed right of individuals to choose their own healers. However, when the case involved infectious disease, contentions often arose over the supposed need to protect the public from contagion. To be sure, in order legally to determine those dangers a public consensus had to form regarding public health dangers. But once that understanding had been determined, the public became less tolerant of an individual's claim to "each his own physician" because that choice would, in the view of the majority, threaten the commonweal.

As we have seen, in the face of this threat to their practice, Christian Scientists effectively lobbied legislators to exempt them from medical licensing laws. Their successes relieved them of the requirements placed on physicians to report contagious disease, but they did not eliminate the growing public fear that practitioners contributed to its spread. To respond to this, Christian Scientists had only the two options represented by Farlow and Baird in 1897—accommodate or defy—and officially they chose to follow Farlow. In several articles and press releases distributed throughout early 1901, Eddy informed the public and her followers that Christian Scientists should report "contagion to the proper authorities when the law so requires. When Jesus was questioned about obeying the human law, he declared: 'Render unto Caesar the things that are Caesar's,' even while you 'render unto God the things that are God's.'"[39] She instructed parents: "Where vaccination is compulsory, let your children be vaccinated, and see that your mind is in such a state that by your prayers vaccination will do the children no harm. So long as Christian Scientists obey the laws, I do not suppose their mental reservations will be thought to matter much."[40]

If Christian Scientists believed this adjustment in their practice would satisfy their opponents, they soon discovered they had miscalculated. A major reason lay in a growing belief that new medical treatments produced wondrous cures for the deadly childhood disease diphtheria. Diphtheria was a frightening, infectious disease that physicians had little effective power to treat before the 1890s. As one historian has described it, "The disease was a desperate affliction, and parents and physicians alike were tortured by the sight of children dying in an agonizing struggle for breath, as the diphtheritic membrane appeared slowly to choke them." To relieve choking, by the mid-1880s physicians were performing tracheotomies, which by the early 1890s were being replaced by intubation. The first real medical rather than surgical breakthroughs against the disease occurred when investigators identified the diphtheria bacillus (*Corynebacterium diphtheriae*) in 1884 and then treated the first person with antitoxin in 1891. When diphtheria antitoxin became generally available after 1894, death rates from diphtheria declined, although it is clear that the decline was not uniform over time or place, and some evidence now indicates it may have been primarily due to the evolution of the bacterium into milder forms.[41]

In the mid-1890s, simultaneous with the wider use of diphtheria antitoxin, the medical press increasingly blamed Christian Scientists for diphtheria's threat to the public's health, and the trials of Christian Scientists involved in the treatment of diphtheria began to take place. As early as 1895, the year of Owen's trial, *JAMA* announced that "the alarming spread of diphtheria" in Indianapolis was "largely due to the practices of 'Christian scientists' who have honeycombed whole districts with their doctrines." The journal warned that the same could apply wherever one found Christian Scientists.[42] Of the thirty-seven pre-1930 Christian Science court cases involving healing practices that I have identified, twelve (32.4 percent) involved a diagnosis by physicians of diphtheria. Each case also involved the treatment and subsequent death of a child (table 2).

In each of the diphtheria trials before 1901 physicians had testified and prosecutors had argued about the dangers of infection to the public as a result of the unsupervised and unquarantined practices of Christian Scientists. But in Toronto, Ontario, in 1901 and in White Plains, New York, in 1902, physicians testified for the first time that proper treatment could have cured the children; in both cases prosecutors filed charges of manslaughter.

In Toronto the Crown charged James H. Lewis with manslaughter for failing to provide proper medical treatment for his young son Roy, who had died of diphtheria. The boy's mother detailed the discovery of the boy's illness and conceded that "the disease was such that if she had been of any other faith she would have called in a doctor to see the child." This admission that the parents had failed to provide the "necessaries of life" for their child laid the foundation for the prosecutor's case. Then Dr. Arthur Jukes Johnson and Dr. J. M. Cotton offered the court "a heap of statistics to show that the rate of mortality in mild cases of diphtheria was very low" and stated that if Roy "had been injected with anti-toxin six days prior to death recovery would have been certain." The jury took only twenty minutes to decide on a guilty verdict, which held up on appeal. That decision, received with satisfaction by physicians, established for Canada that "the law of the land demands that medical science shall be consulted in cases of children."[43]

Less than one year later the diphtheria-related death of seven-year-old Esther Quimby in White Plains, New York, presented United States

Table 2. Diphtheria and the Trials of Christian Scientists

Date	Charge	Child	Death	Location	Name
1893	Unlicensed practice	X	X	Beatrice, Nebraska	Buswell
1895	Failure to report	X	X	Davenport, Iowa	Owen
1895	Manslaughter	X	X	Toronto, Ontario	Beer
1897	Failure to report	X	X	Kansas City, Missouri	Baird
1897	Unlicensed practice	X	X	Washington, D.C.	Sessford
1899	Unlicensed practice	X	X	Minneapolis, Minnesota	Brookins and Meyer
1900	Unlicensed practice	X	X	Milwaukee, Wisconsin	Arries and Nichols
1901	Manslaughter	X	X	Toronto, Ontario	Lewis
1902	Manslaughter	X	X	White Plains, New York	Quimbys and Lathrop
1902	Child neglect	X	X	Los Angeles, California	Reeds
1920	Manslaughter	X	X	Newark, New Jersey	Walkers
1925	Manslaughter	X	X	Manitoba, Canada	Watsons

Note: 1884, Diphtheria bacillus identified (*Corynebacterium diphtheriae*); 1891, first person treated with diphtheria antitoxin; 1894, diphtheria antitoxin first generally employed.

authorities with the opportunity to establish a similar legal precedent in their country. For reasons I will explore closely below, prosecutors failed to convict her parents and her practitioner of manslaughter by neglect. Nonetheless, in great part owing to a changed public attitude toward the effectiveness of medical treatment for diphtheria, the Westchester County grand jury, which had failed to indict in similar circumstances on three previous occasions, indicted John Carroll Lathrop, John Quimby, and Georgiana Quimby for "maliciously and feloniously" causing Esther's death "by neglecting to provide medical attendance."[44]

The medical community had convinced the grand jury that it could consistently diagnose diphtheria, that it possessed a correct understanding of its etiology, and that it offered an effective treatment. No longer could Scientists convincingly argue, as they had before, that physicians just wanted a monopoly or that reasonable people could differ about the most effective treatment for diphtheria. The public's acceptance of the authority of scientific medicine over this infectious disease meant that the argument for freedom of therapeutic choice, at least for children, had been blunted. Since Scientists still believed they should be allowed to choose Christian Science treatment for their diphtheria-infected children, they would have to defend their actions on different legal grounds.

Still reeling from the Lewis conviction in Canada and anxious about the prospects in White Plains, Christian Scientists learned in October 1902 that "the Massachusetts State Board of Registration in Medicine has decreed that hereafter no Christian Science healer will be allowed to treat any patient suffering from a contagious or infectious disease. This will also apply to all other forms of treatment not recognized by law."[45] Continuing down the path of discretion she and Farlow had chosen earlier, Eddy advised that "until the public thought becomes better acquainted with Christian Science, that Christian Scientists decline to doctor infectious or contagious diseases." In fact, she continued, "Christian Scientists should be influenced by their own judgment in the taking of a case of malignant disease, they should consider well as to their ability to cope with the case—and not overlook the fact that there are those lying in wait to catch them in their sayings; neither should they forget that, in their practice, whether successful or not, *they are not specially protected by law.*"[46]

Physicians greeted this news with glee and more than a little self-satisfaction. They congratulated themselves that "the storm of protest from physicians and health commissioners has had a very decided influence, and hereafter we may expect Christian Scientists to pay closer heed to the demands of public health protection."[47] But they also observed that Eddy's decision appeared to be a "clear sacrifice of her essential first principles as to the absolute non-existence of disease."[48] "If contagious and infectious diseases are admitted," opined the *Northwestern Lancet*, then "the cult must also admit that there is such a thing as diagnosis, and if they would investigate the 'false belief' still further they would soon discover how difficult it is to differentiate between the various forms of contagion."[49] Testimony taken at the Esther Quimby inquest made it clear that the Christian Scientists had faltered on precisely this issue. Whereas Lathrop had treated Esther for tonsillitis, her father, John Quimby, had asked a physician to report it to the board of health as diphtheria.[50] Given this apparently fatal contradiction between a rejection of the knowledge necessary for diagnosis and the willingness to diagnose and report diseases, many physicians hoped with *JAMA* that this was "the beginning of the end of the Eddyite delusion."[51] However, more had been at stake in these two cases than the proper treatment for diphtheria; the very right of Christian Scientists to care for their children according to their religious beliefs had been in the balance.

Children, Medical Treatment, and Parental Responsibility

After the Civil War, many Americans grew increasingly concerned about society's treatment of children, and activists launched efforts to secure their protection.[52] In New York City in 1873 a group of reformers formed the first Society for the Prevention of Cruelty to Children (SPCC), one of many such societies across the country that worked to ensure their safety and well-being. Societies such as the SPCC and numerous other private and public organizations worked at the forefront of a Progressive Era movement whose efforts led to the passage of scores of child protection laws in the areas of labor, education, control of vice, and health.[53] By 1889 the Brooklyn, New York, SPCC had already recommended action to prevent the spread of childhood diseases by the honest yet dangerous activities of religious healers.[54]

The goals of these child welfare reformers dovetailed neatly with the growing concern among some physicians for the health of children and led to the formation of organizations such as the American Association for the Study and Prevention of Infant Mortality (1908; renamed American Child Hygiene Association), the Child Health Organization (1917), and in part, to the formation of pediatrics as a medical specialty.[55] However, under the leadership of physicians such as Abraham Jacobi, the founder of the American Medical Association's Section on Diseases of Children (1880) and the first president of the American Pediatric Society (1888), pediatrics became rooted as well in a scientific recognition of the distinct anatomy and physiology of children and of their often unique diseases and health needs.[56] The pediatricians who emerged from these developments after the turn of the century constituted a core of vocal and influential advocates for the scientific health care of American children and for their protection from poor public health practices, "quacks," and well-meaning but "misguided" parents.

Simultaneously, adjustments occurred in the ways America's legal system dealt with children and oversaw the changing standards of child care and family life. A juvenile court system arose near the turn of the century as one of the new social institutions responsible for ensuring a family met its obligations to its children. According to historian Michael Grossberg, these new legal measures "were designed to save families," not to destroy them. "They sought to keep a child from being taken from his home by compelling parents to provide suitable education, clothing, food, and moral instruction." Such innovations also "testified to a dawning concept by legal authorities of children as separate, if naturally dependent, individuals with their own needs and interests."[57]

American developments in these three areas—child welfare, the health care of children, and juvenile law—intersected in the early 1900s over the issue of whether religious healing adequately protected children's health and safety. Of the thirty-six pre-1921 Christian Science cases involving healing practices that I have identified, twenty-one (58.3 percent) involved a sick child, and in at least sixteen (44.4 percent) the death of a child precipitated the trial. Moreover, of the ten cases in which the state charged practitioners or parents with manslaughter, eight (80 percent) involved the death of a child. Although the deaths of Christian Science children had triggered court trials since the 1880s, the turn of the century witnessed the appearance of a new legal tactic to address them: the indictment and trial of parents. After 1900, as table 3 clearly shows, prosecutors targeted Christian Science parents, not just practitioners, for prosecution when their children died from an infectious disease that was not medically treated.

In a precedent-setting 1899 English case (*Queen v. Senior*) involving a religious group called the "Peculiar People," who practiced faith healing and shunned physicians, Chief Justice Lord Russell affirmed a father's manslaughter conviction in the death of his eight-month-old infant. The judge, having determined that the father had broken England's Prevention of Cruelty to Children Act of 1894, concluded: "At the present day, when medical aid is within the reach of the humblest and poorest members of the community, it cannot reasonably be suggested that the omission to provide medical aid for a dying child does not amount to neglect."[58]

Thus, when Canadian Chief Justice Glenholme Falconbridge presided over the Lewis case in 1901, he possessed not only a Canadian criminal code that required parents to provide "necessaries for any child,"

Table 3. Deaths of Children and the Trials of Christian Scientists

Date	Infectious Disease	Child's Death	Parent Charged	Practitioner Charged	Name
1887		?		X	Bunnell
1888		X		X	Corner
1889	?	?		X	Stewarts
1893	X	X		X	Buswell
1895	X	X		X	Hatten
1895	X	X		X	Owen
1895	X	X		X	Beer
1895		?	X	X	Samis and Cook
1896		?		X	Chester
1897	X	X		X	Baird
1897	X	X		X	Sessford
1899	X	X		X	Brookins and Meyer
1900	X	X		X	Arries and Nichols
1901	X	X	X		Lewis
1902	X	X	X	X	Quimbys and Lathrop
1902	X	X	X		Reeds
1903	X	X		X	Marble
1907	X	X	X		Watsons
1907	X	X	X		Byrne
1911				X	Cole
1920	X	X	X		Walkers
1967	X	X	X		Sheridan
1984	X	X	X		Walker
1984	X	X	X		Glasers
1984	X	X	X		Rippberger/ Middleton
1988		X	X		Kings
1988		X	X		Twitchells
1989		X	X		Hermansons
1989		X	X		McKown

but also Lord Russell's legal precedent that added medicine to the required necessities of food and clothing.[59] Consistent with the English court, Falconbridge instructed the jury that "every man has a right to employ such a doctor or surgeon as he pleases or to employ none, but in the case of a child it is different, and here is where the law steps in and draws the line."[60] The jury's subsequent conviction of Lewis affirmed the responsibility of Canadian parents to provide medical care for their children and reflected a growing confidence among Canadians in the efficacy of scientific medicine. But for Christian Scientists the decision not only forced them into unprecedented levels of cooperation with scientific medicine in Canada but also underscored the importance of gaining favorable decisions in United States courts if their movement was to sustain its persistent rejection of medicine. Would Americans affirm the decisions made by England and Canada to require medical treatment for their children? And if so, given their earlier fight against being regulated as medical practitioners, would Christian Scientists be allowed to continue administering Christian Science treatments to their children?

On 22 October 1902 Archibald T. Banning, coroner of Westchester County, New York, announced to the press that an autopsy had revealed that the daughter of Mr. and Mrs. John Quimby died of "diphtheria and Christian Science neglect."[61] "In this case," the coroner continued, "the child was only 7 years old, and although I understand her parents claim that she was a 'true Scientist,' their claim will not hold legally, as a child in New York State is not responsible for her own acts until she is 14."[62] John Carroll Lathrop, the practitioner who had attended the child, met reporters later that evening to express his heartfelt regrets "that the child was not cured, but at the same time I think it is very well to remember that Christian Scientists are not the only people who occasionally lose a patient.'" William D. McCracken, the Committee on Publication for the state of New York, agreed with Lathrop. He believed that "the mere fact that a method of healing the sick is chosen by the parents which is not in accordance with the prevailing notion does not suffice to establish the charge of neglect, else there could be no advance in the art of healing. Even should a failure or casualty result after Christian Science treatment has been given, this occurrence cannot be taken as evidence of neglect on the part of the parents, for such failures occur daily in great numbers under the most approved treatments of materia medica."[63]

John Lathrop's mother, Laura Lathrop, a well-known Christian Science pioneer in New York City, noted that "Christian Scientists have been held for the Grand Jury before and nothing has ever come of it. We are not aggressive, but we are not worried in the least."[64] Nonetheless, in case they had to go to trial, Scientists gathered funds for a war chest and launched a public relations blitz. As John Lathrop put it, "We do not think publicity does us any harm"; in fact, "the more that is printed about us a greater interest is taken in our religion."[65]

As it turned out, they needed all the help with the public that they could muster. Anxious parents complained about unreported contagion in their neighborhoods and schools, and area clergymen excoriated Christian Science and its founder. Even the membership of the grand jury appeared to be stacked against the Christian Scientists, since the panel comprised "Episcopalians, Baptists and Presbyterians" who were "opposed to the practice of Christian Science."[66] Undoubtedly, editorialized the *New York Evening Journal*, "the Christian Scientists are good people, honest, earnest, conscientious people. BUT CHRISTIAN SCIENTISTS MUST NOT MURDER CHILDREN." This child

> was allowed to die, and it did die, while a so-called faith-curer, or alleged "healer," prayed and went through various incantations as senseless, as hopeless, and as vicious in a case of diphtheria as would be the contortions of some negress of the voodoo faith twisting serpents around her neck to achieve a medical result. . . . All forms of religious belief are respectable and should be respected, as long as they confine themselves to predictions of events which come after this life, and as long as they do not interfere with the welfare of those now living. But when any religion threatens the moral health, as does polygamy, those who advocate it should be punished. And when any religion threatens the physical health, those responsible for physical injury should be punished. . . . An example is needed now. If any alleged healer, pretending to cure that child, deprived her of medical aid, and without legal license pretended to practise the healing art, he should be convicted of manslaughter and put in jail for a term sufficiently long to act as a preventive to him and A WARNING TO OTHERS.[67]

Apparently the grand jury agreed, because on 31 October it indicted all three defendants on charges of manslaughter in the second degree.

Less than a month later, when defense attorneys filed demurrers on behalf of their clients, they argued that "every citizen was entitled to full liberty in his religious belief; that there was no law that compelled a Christian Scientist to call in a regular practitioner in case of sickness, and because of the non-existence of such a law the indictment was faulty."[68] We have met both arguments several times by now, and the argument for religious liberty is easily recognizable. However, the argument that no law compels Christian Scientists "to call in a regular practitioner in case of sickness," is much more ambiguous. If the defense intended to emphasize "Christian Scientist," then it meant that Christian Scientists were an exception to any general rule that required parents to consult physicians for their sick children or run the risk of being charged with neglect. However, no such exemption existed in either child welfare or medical practice law; furthermore, the context of the case makes it more likely that the defense meant to emphasize "regular practitioner," as the following discussion makes clear.

Ever since the Westchester County coroner announced his inquest, official spokespersons for Christian Science had echoed the sentiments of head COP Alfred Farlow: "We trust that while the New York Coroner is calling a Christian Scientist to account for the loss of one diphther[i]tic case, he will at the same time account for the 1,100 fatalities of the same character which the records show have occurred in Greater New York during the past six months."[69]

Christian Scientists believed that the deaths by diphtheria of so many persons, most under the care of a physician, proved the ineffectiveness of scientific medicine and illustrated the unreasonableness of requiring parents to take their children to regular physicians. True, they had agreed to suspend the treatment of contagious diseases for a time, but that did not mean they had capitulated on the underlying issues. Christian Scientists still meant to challenge the implication of Canada's Lewis decision and the White Plains grand jury indictment that, at least with regard to diphtheria, "scientific medicine" had begun to replace the "schools of medicine" in the minds of the public. The demurrers implicitly granted that the state could require medical care for children, but they challenged its right to mandate care by a "regular physician."[70]

The court delayed action on the demurrers for nearly three years while it awaited the outcome of the appeals of White Plains Dowieite J. Luther Pierson, convicted 21 May 1901 of failing to provide medical attendance to his daughter. In a closely watched case supported both morally and financially by Dowieites and Christian Scientists, Pierson's attorney, Robert E. Farley, successfully argued before New York's appellate court that "if there were a law of this State compelling one to furnish medicine to his infant child, it would be unconstitutional, and with greater force can this be urged as to a law compelling one to furnish medical attendance."[71] However, in October 1903 a divided state court of appeals overturned the appellate court decision and affirmed the original conviction on the grounds that the state's role as *parens patriae* superseded Pierson's right to practice his religion.[72] Both physicians and lawyers greeted the news with pleasure, and the *New York Times* concluded that "a court of last resort has in unmistakable terms declared that the profession of a belief in faith cure does not constitute a valid defense against the charge of refusing to employ a licensed physician in cases of dangerous illness."[73] Similar decisions regarding faith healing in Pennsylvania (1903), Indiana (1904), Maine (1905), Georgia (1908), and Oklahoma (1911) followed in quick succession.[74]

Still, it was unclear whether the law viewed Christian Science healing as "faith cure" or as "medical practice" and, if the latter, whether it was acceptable treatment for children. The 1902 Los Angeles case of Merrill and Clara Reed, charged with willfully neglecting their diphtheritic child, did not clarify the issue. Rather than ruling on the action of the parents, the jury looked at their intent; it "stuck at that word 'willfully'" and acquitted them.[75] Similarly, county justice William P. Platt finally granted the demurrers of Lathrop and the Quimbys on 8 August 1905.[76] But in 1907 Clarence W. Byrne, a Christian Scientist in New York City, received a sentence of thirty days in the penitentiary for failing to provide medical attendance to his six-year old daughter, making it abundantly clear that there was no guarantee that juries or judges even in New York would follow the White Plains decision.[77] Clearly the courts had moved toward the opinion that parents broke the law when they did not provide medical care for their children. However, courts remained ambivalent about whether Christian Science treatment constituted medical treatment; and if it did, courts balked at convicting parents for practicing their religion in what they had believed to be their children's best interest.[78]

Christian Scientists learned the hard way that, in the words of Eddy, "to teach and to demonstrate Christian Science before the minds of the people are prepared for it, and when the laws are against it, is fraught with danger."[79] Fearing a resurgence of anti–Christian Science legislative activities, the editor of the *Christian Science Sentinel*, Archibald McLellan, urged his readers in January 1906 to give "particular attention . . . to the systematic distribution of the authorized literature of our denomination." In concert the Christian Science Publishing Society issued a new pamphlet, *Christian Science and Legislation*, designed not only so that the "opponents of Christian Science may have an opportunity to acquire correct information upon this important subject, but also that Christian Scientists themselves may know what their rights are."[80] And in yet another accommodation to the concerns of public health officials, Eddy ordered that "if a member of the Mother Church shall decease suddenly, without previous injury or illness, and the cause thereof be unknown, an autopsy shall be made by qualified experts."[81]

Possibly more than any other public events, the court trials of Progressive Era practitioners influenced the evolution of Christian Science and hastened the shift from its medical roots to its religious foundation and from a nonconformist to an accommodationist attitude. Christian Scientists had adapted their therapeutic system to the demands of the state while retaining their right to believe and practice their religion. But these trials also profoundly affected American attitudes toward religious healing. Medical licensing and public health laws slowly evolved to accommodate religious healing. States exempted practitioners from examinations in anatomy, physiology, or materia medica, but they compelled them to report births and deaths and to abide by the rules of notification and quarantine for contagious diseases. However, in the cases of children and persons of diminished mental capacities, Americans still struggled through their courts and legislatures to balance the responsibility of the state to protect its citizens against the right of parents to care for their children according to their honestly held religious beliefs.

8

Century of Promise, Then Peril

The American Academy of Pediatrics recommends that all pediatricians, pediatric surgeons, and AAP state chapters vigorously take the lead to (1) increase public awareness of the hazards to children growing out of religious exemptions to child abuse and neglect legislation; (2) support legislation in each state legislature to correct statutes and regulations that permit harm to children under the shield of religious exemption; (3) work with other child advocacy organizations and agencies to develop coordinated and concerted public and professional actions for recision of religious exemptions.
—Committee on Bioethics, American Academy of Pediatrics, 1988

As the 21st century approaches, is religion simply a kind of obsolete cultural window dressing, to be tolerated only insofar as it isn't taken too seriously? Are religious differences permissible only on matters that make no practical difference? The movement to repress religious healing . . . seeks to mete out punishment "to those who have practical expectations from religion."
—*Christian Science Monitor*, 1989

When pneumonia ended Mary Baker Eddy's long and productive life in December 1910, she left devoted followers, an apparently stable organization, and a pragmatic legacy of accommodating Christian Science to the changing winds of medical legislation and the surging public support for efforts to control infectious disease.[1] Still strongly deferential to Eddy's vision, twentieth-century Christian Scientists set out to discover how well Science and the Eddy corpus fitted them to solve institutional and public crises and to meet future challenges. But because Eddy's prophetic authority had been so essential to settling disputes, affecting doctrinal change, and updating practices, her death presented Scientists with a dilemma. How could the church sustain its dependence on the authoritative writings of Eddy while maintaining the flexibility necessary to adapt to new circumstances? If Scientists chose to sustain their allegiance by requiring absolute obedience to a literal reading of Eddy's writings, they risked becoming irrelevant to a changing world. If they followed her example of adaptation and change, who possessed the authority to exact obedience to new interpretations or doctrines?[2]

The dilemma became a crisis in 1916, when the board of directors of the Mother Church and the Board of Trustees of the Christian Science Publishing Society became enmeshed in a struggle over the right to define orthodox Christian Science practice and belief. With each group claiming that Eddy had given it the right to pass upon the doctrinal orthodoxy of church publications, the debate grew acrimonious and finally found its way to the courts of Massachusetts for resolution. The 1922 decision of the courts, supported by a majority of church members, sustained the claims of the board of directors and allowed it to cautiously adapt Eddy's teachings to the modern world.[3] Such public disputes did not immediately seem to erode church membership—the United States Religious Census of 1936 reported 268,915 American Christian Scientists. But as the century progressed and churches closed, the number of practitioners decreased, congregations grayed, and the criminal prosecution of Scientist parents resurfaced, it became obvious that the appeal of Christian Science for Americans had significantly declined.[4]

Living "Science" in the Twentieth Century

After the 1922 court decision, esteem grew among the faithful for the Founder or Leader, as Christian Scientists often called Eddy, and for *Science and Health*. Healing, broadly defined by the board of directors' influential *Century of Christian Science Healing* as "the rescue of men from all that would separate them from the fullness of being," remained the core of twentieth-century Christian Science.[5] However, while still insisting that healing could never be separated from a true understanding of the nature of reality, Scientists more readily acknowledged that gaining such understanding often was painfully slow and more freely conceded the use of physical aids. While still avoiding the use of physicians and psychologists, Scientists used hearing aids, accepted blood transfusions, employed obstetricians, permitted physicians to set broken bones, and consulted with practitioners "on the anatomy involved" in complicated cases.[6] When Arthur Nudelman studied the behavior of Christian Science university students in the late 1960s, he discovered that on the whole they appeared to make little use of medical facilities; however, they wore eyeglasses when necessary, visited dentists, and generally utilized "medical services more frequently for mechanical problems than for ailments of other types."[7] Eddy had encouraged similar accommodations during her lifetime, but in a further effort to justify such "mingling with the material," some Christian Scientists stated that physicians had become more Scientific by holistically treating mind, body, and soul. However, if practitioners received medical care for their own illnesses, church officials removed their names from approved practitioner listings for a time.[8]

Among the errors of belief that Scientists struggled against, death still presented the greatest challenge. Not only did the death of a loved one create profound grief, but for Scientists death strongly challenged faith by presenting compelling evidence that Christian Science treatment had failed. Confronted by the "error" of death, many found it hard to affirm with Christian Science historian Robert Peel that death is only "one phase, however grievous it may seem, of men's present imperfect sense of life." To assist them in their attempt, most Scientists emphasized death's transitory nature, referred to it as "passing on," and humbled themselves by remembering that the truth of Science had not yet completely overturned the errors of human existence. Although most Christian Scientists probably held funeral services for the departed, the casket remained closed (if the body had not been cremated) during a quiet service of readings from the Bible and *Science and Health*.[9]

A Century of Christian Science Healing, published in 1966, extolled physical healing as "one of the most concrete proofs that can be offered of the substantiality of Spirit," and Scientists celebrated their ties to Christian healers in all ages. Christian Science metaphysics, they believed, had even led modern medicine to moderate its emphasis on the body as machine.[10] Beginning in the mid-1920s, Scientists increasingly emphasized emotional and spiritual harmony, not simply physical well-being, as the highest expression of healing, thereby explicitly countering any claim that psychiatry and psychology produced results that Christian Science could not surpass. And exercising a new ecumenical generosity, Scientists even found important similarities between their own mission to heal the sick in Christ's name and the motivations that stimulated the charismatic movements of many late twentieth-century denominations.[11]

In many respects Scientists maintained a fairly conservative lifestyle that included ordinary attention to wise choices about food, housing, personal hygiene, and sanitation. They avoided so-called vices, believing with Alan A. Aylwin, former associate editor of the *Journal*, that Christian Science can show "the way of escape from all forms of addiction, whether to heroin, tobacco, alcohol, caffeine, masturbation, or just plain overeating."[12] Scientists abstained from alcohol and tobacco not for physical reasons but because their addictive powers could compromise mental freedom.[13] Despite such healthful practices, however, a coroner's study in the early 1950s revealed that American Christian Scientists experienced higher death rates from malignancies and heart disease than the national average and did not live as long on the average as the general population, findings confirmed by a study published in *JAMA* in 1989.[14]

For many Christian Scientists the practitioner or "healing worker" remained the principal position in the Christian Science church, although their numbers steadily dropped from an estimated ten thousand worldwide in the 1940s and 1950s. Working as "something of a combined minister and doctor," practitioners in the late 1980s charged between $7 and $15 a day for their treatments, but they often reduced fees if the patient responded slowly.[15] Striving to make their work relevant to a modern world, practitioners emphasized the completeness and satisfaction of living in harmony with the principles of Christian Science and turned their essentially unchanged healing techniques to such modern "problems" as homosexuality, drug addiction, obesity, marital troubles, and job tensions.

Extending their defense of Prohibition in the 1920s and their post–World War II relief efforts, twentieth-century Christian Scientists embraced a more active social ethic and expanded their aid to and treatment of such global problems as natural disasters, political tensions, and social crises. The Mother Church often raised and distributed money, food, or clothing to alleviate the immediate physical consequences of famine, war, unemployment, racial discrimination, and natural disaster—then turned to the decisive mental work that would correct the human misunderstanding of reality that lay at the root of such evils.

A four-part series in the November 1984 *Monitor* titled "Hunger in

Africa" well illustrated this blend of social responsibility and moral obligation. After presenting a fair and accurate assessment of the political, cultural, and climatic problems contributing to starvation in East Africa, in their concluding article, "What Can You Do to Help?" the authors listed the names and addresses of fourteen aid and relief agencies that would accept contributions.[16] However, the author of the *Monitor*'s regular "Religious Article," which began the series, told readers that a correct understanding of reality would provide the best solution for African starvation and claimed that "we can help those who may be thousands of miles from us, not only through charitable giving but through potent prayer."[17]

Christian Scientists sought to convince the world that despite their belief in the unreality of evil, they remained sensitive to human suffering. One such effort found expression in the establishment of two sanatoriums, one near Boston and the other in San Francisco, under the guidance of the Christian Science Benevolent Association (1916). These institutions provided asylum for patients who wished to receive treatment and nursing care away from the ordinary pressures of life, but they did little to convince critics of their compassion.[18] While some took the sanatoriums as evidence that Christian Scientists were learning how to deal with diseases that needed to be isolated, critics viewed the establishment of these resorts for the *"so-called* sick" as simply a place for Christian Scientists to "get so-called sicker" until they are "so-called dead" and a "so-called undertaker takes them to a so-called cemetery," where "they remain with the vast majority silently awaiting the blessed hope."[19]

Approved by Eddy in 1908, formalized nursing care provided further evidence of the compassion of Christian Scientists. Seen as a clearly subordinate but increasingly popular alternative to a career as a practitioner, nursing grew throughout the twentieth century; the July 2000 *Journal* listed 460 certified nurses in the United States alone. Christian Science nurses cared for the physical needs of their patients and assisted practitioners by thinking pure thoughts that made a positive contribution to a healing atmosphere. In a *Journal* article in 1979 Marco Frances Farley succinctly summarized the duties of a present-day nurse: "The nurse dresses wounds and keeps the body clean, comfortable, and nourished so that it intrudes less on the patient's thought."[20] Parallel to earlier developments between physicians and nurses, practitioners often "cured" patients, whereas nurses "cared for" them. During the last decades of the twentieth century the church standardized nursing education to include classroom instruction at accredited nurses' training schools and practical training at several Christian Science sanatoriums and scores of nursing homes. Through these efforts Scientists endeavored to improve the quality of all their nurses, including private-duty nurses, sanatorium nurses, and visiting nurses, and to ensure standards that aided compliance with Medicare and Medicaid programs after 1965.[21]

Before it closed in 1975, Pleasant View Home for the Aged in Concord, New Hampshire, provided an answer to critics who asked "Why doesn't your own church care for its elderly people?"[22] Pleasant View's pioneering work provided a model for numerous other nursing homes and retirement facilities that continued to create an environment of compassion and care in which Scientists could prove that they wanted to assist "humanity in meeting and mastering the needless limitations erroneously associated with advancing years."[23]

Lobbying for Legislative Protection

The fight in New York courts over the Cole case, which had begun in 1911, ended in 1916 with the pivotal decision by the state court of appeals to affirm the legal right of Christian Scientists to both believe and practice their religion. But the divisive process underscored the culturally ambiguous nature of Christian Science—both religion and medicine—in a twentieth-century society enamored of biomedicine and increasingly anxious about medical pluralism, and it virtually guaranteed Scientists a continued struggle against medical legislation and for public acceptance. Scientists, however, did not stand alone in their opposition to M.D.s; other groups also decried the supposed trend toward a "monopoly" in American health care. To continue the fight against the medical legislation that had so bedeviled them for decades, many Scientists believed they should discreetly unite forces with those like-minded groups.

An estimated 60 to 75 percent of the 200,000 church members joined and partially funded a 1910 coalition called the National League for Medical Freedom, comprising homeopaths, eclectics, osteopaths, druggists, antivaccinationists, and antivivisectionists, among others.[24] Members of the League believed that "political doctors have adopted the mask of 'the public welfare,' 'public health,' 'sanitation,' etc., to secure the confidence of the people," and therefore members needed "to disseminate information pertaining to, and to safeguard through education and publicity, the rights of the American people against unnecessary, unjust, oppressive, paternal and un-American laws ostensibly related to the subject of health."[25] The League gained national prominence when it engaged in a successful campaign in 1912 to derail the formation of a national department of health. United States Senator John D. Works of California, a Christian Scientist, spoke for two days before the Senate in opposition to the initiative, which he believed would place American medical practice under the dominance of the American Medical Association and lead to the oppression of Christian Scientists by federal health officers.[26]

Just five years later, California's Christian Scientists joined the battle against compulsory health insurance in their state. Peter V. Ross, the COP for northern California, and Henry Van Arsdale, his counterpart in the south, marshaled the state's Christian Science membership in a publicity campaign that ironically put them on the same side of the issue as many California physicians. Like others opposed to state-mandated health insurance, the Scientists raised the German bogeyman, decried the "force and coercion" at the heart of the proposed system, and touted American "ideas" and "sacred rights."[27] But they also argued against a plan that they believed supported "exclusively a material method of healing in preference to a spiritual method." Partially owing to these efforts by Christian Scientists and their political influence on the governor, Californians rejected compulsory health insurance.[28]

Despite cooperative efforts such as these, tensions and distrust generally still dominated the relationship between physicians and Scientists. By scrupulously reporting every incident of contagion associated with a Christian Scientist and challenging most legislative moves the COPs made, physicians exposed their fears that Christian Scientists still posed a threat to public health. But as physicians grew aware of an expanding public confidence in scientific medicine, they became more sure of their position and less concerned with Christian Scientists as rival healers. When yet another bill to exempt Christian Scientists from medical practice law came before the Ohio legislature in 1919, the Ohio State Medical Association took no position. The chairman of the association's legislative committee simply "pointed out that the proposed measure would not in any way affect physicians but that it was a menace to the public health, in permitting unqualified and untrained persons to treat the sick. As this is a matter which concerns the health of the people of the state rather than the welfare of physicians, it is only right that it should be settled by the people themselves."[29] When the bill failed eighty-three to twenty-two, *JAMA* speculated that organized medicine had fared "far better than would have been the case if its members had participated aggressively in political activities against the measure."[30] L. W. Zwisohn, a physician from New York, agreed: "During the last thirty years while going to the legislature opposing bills of the osteopaths, Christian scientists and the like, I came to the conclusion that the best way to deal with them is to ignore them."[31] That may have been true, but many physicians still recognized Christian Scientists as formidable opponents who maintained "the most efficient and best-oiled political lobby to be found anywhere."[32]

While physicians began to relax in the luxury of their newfound authority, Christian Scientists remained vigilant for infringements of their reputed rights. They repeatedly announced that Christian Scientists had "been more scrupulous than most people in obeying laws which require vaccination, the reporting or isolating of contagious diseases, and acts of a similar character."[33] And like New York City Scientist Albert F. Gilmore, they wondered aloud why the state refused to indict physicians whose diphtheria patients died or why the state should be allowed to destroy the "sanctity of the home" by interfering with parents who treated their children by Christian Science.[34]

During the twentieth century, Christian Science COPs responded to medical legislation and official health regulations according to the principles of "exemption" and "inclusion." Ostensibly the committees took no action against "sanitation, quarantine, and general public health and safety requirements," although they opposed the fluoridation of public water supplies during the late 1950s and early 1960s based on "individual freedom and rights of conscience."[35] In 1949 Ohio, which had first forced practitioners to comply with medical practice laws in 1915, became the last state to exempt them. And whenever a law to enforce the use of medical or psychiatric care, to require compulsory vaccination for schoolchildren, or to impose compulsory physical examinations came before state legislators, then the ever vigilant COPs pursued exemptions for Christian Scientists.

By the early 1980s legislators in all states except West Virginia and Mississippi granted religious exemptions to state laws that required the immunization of schoolchildren, but outbreaks of contagious diseases traceable to such exemptions had proved difficult for public health authorities to deal with. From 1951 to 1954 Cora Sutherland, a California schoolteacher under treatment by a practitioner for tuberculosis but exempt from a required chest X ray, exposed hundreds of students to her disease before she died, and in 1972 poliomyelitis ravaged a Christian Science school in Connecticut and left eleven paralyzed.[36] Measles, however, proved the most persistent problem. Ten years after the Centers for Disease Control announced in 1978 its program to eliminate measles, the number of reported cases had declined by more than 70 percent. But the disease persisted among religiously exempt populations. A recurring problem for the residents of Principia, a Christian Science college near St. Louis, measles swept the campus in 1985 and again in 1994. Deaths and quarantine followed the 1985 outbreak, and in response to both episodes many students and staff submitted to vaccination.[37]

Children, Medical Care, and the Law

Beyond seeking legislative protection through exclusionary clauses in the law, twentieth-century Scientists sought legislative "inclusion" to ensure that the law considered "Christian Science treatment and care" as an "alternative to medicine." Given their ambivalent legal status after the conflicted judicial decisions of the first decade of the century (see appendix), they believed such legislative provisions would secure the right of Christian Science parents to care for their children's medical needs according to their religious beliefs.[38]

As early twentieth-century Christian Scientists, physicians, and the public debated public health legislation and the safety of children, they repeatedly returned to the issue of medical efficacy. They argued about who had done a better job during the 1918–19 influenza pandemic, and into the 1920s they traded statistics about the effect of Christian Science versus antitoxin on diphtheria.[39] In those debates, however, Christian Scientists walked a tightrope. They believed Christian Science healing was the practice of religion, but they wanted to make sure they were not seen to be like Dowieites and other faith healers who rejected medical care for their children. They still wanted, in the earlier words of head COP Alfred Farlow, "to establish the healing efficacy of Christian Science and a reasonable excuse for the non-use of medicine on the part of Christian Scientists by showing that a child under Christian Science treatment has resources and means at hand for the maintenance and restoration of health equal if not superior to medicine."[40] In short, Christian Science healing was religious healing that provided medical care of a higher order.

But the American legal system struggled to juggle its own set of competing interests. Children got sick and died all the time. When did a parent's decision not to consult a physician become a prosecutable offense? When parents consulted a religious healer, did religious freedom protect them? And how did the death of an ill child justify the prosecution of loving and responsible parents? As the *New York Times* resignedly editorialized on the occasion of the 1920 New Jersey manslaughter trial of Mr. and Mrs. Andrew Walker after their nine-year-old daughter died of diphtheria:

> One father or mother calls in the doctor for every little ailment or indisposition that can be seen or imagined in a child; others, no less loving or well-intentioned or intelligent, are not so easily alarmed and trust longer to domestic knowledge and skill.
>
> Somewhere along this line criminality begins, but just where it is hard to say—so hard that Judges and jurors refuse, if they possibly can find excuse, to declare that the limits of proper discretion have been passed.
>
> And hence the rarity of Christian Science convictions and the frequency of what are vaunted as Christian Science vindications. Never is the evidence of neglect as strong in the case of other maladies as in that of diphtheria, which accounts, of course, for the fact that most of the prosecutions follow deaths from that cause. Even then, however, acquittals are almost invariable. The element of ill intention is absent, the right of free exercise of religion is involved, or rather is invoked, and so, though the wrong is obvious, the law, in practice, can neither deter nor punish.[41]

Christian Scientists recognized that in such a climate of conflicted public opinion, a safer haven might be found in embracing religious liberty than in claiming therapeutic superiority. At the 1926 annual meeting of the Mother Church, Judge Clifford P. Smith, the church's COP, filed the following report:

> All members of our church can also help to guard our legal rights by comprehending them definitely, and by distinguishing between those which

are more important and those which are less so. Of first importance is our right to practice our religion and to depend upon its practice, which right includes that of employing the aid of a practitioner. Also of importance is our right to be free from unnecessary and unreasonable regulations and requirements pertaining to health or religion. At this point we come in contact with other people, who may not distinguish between health and medicine, and whose opinions as to what is necessary and reasonable may differ from ours. Of course, our giving due consideration to them, as our religion teaches, will help us to comprehend and maintain our own rights, and will keep us from exaggerating the importance of requirements which are merely irksome.[42]

One senses in these remarks that Smith realized that Christian Science had failed to become the American standard for medical treatment, and that the continued sacrifices of simpleminded martyrs might not turn back the clock. Instead, to remain relevant and avoid complete eradication, Christian Scientists needed to separate the essential from the peripheral in their practices. To that end Smith suggested that the most important right for Christian Scientists was the right to practice their religion by healing within their own community. Of secondary importance was the right to be free of any "unnecessary and unreasonable" restraints on that practice, but in this regard Smith recognized that many Americans now believed that the unrestrained practice of Christian Science endangered the public. Thus, in an affirmation of Eddy's long-standing policy, Smith suggested that by willingly giving up their right to unrestrained practice and by accommodating it to the secular community, Scientists would be most true to the core of their beliefs.

This approach seemed to work well during the decades of the Great Depression and World War II. Even when prosecutors in 1956 indicted Mr. and Mrs. Edward Cornelius, Christian Scientists living near Philadelphia, for involuntary manslaughter after their infant son died from diabetes, the policy of accommodation proved to be their salvation. At the urging of the COP and others, the Pennsylvania legislature had legalized spiritual healing. And since 1920 the state's courts had contributed to the public legitimacy of Christian Science by granting charters to churches, sanatoriums, and nursing services. By implication, such state actions freed Christian Scientists to pursue their religious activities, which included healing. The prosecutor, acknowledging such "'present day legislative acts'" and concluding that "'diabetes did not subject the public to any danger,'" moved to withdraw the indictment, and the court concurred.[43]

From the 1920s through the mid-1960s, other prosecutors undoubtedly confronted the decision whether to charge parents and practitioners in the deaths of Christian Science children. Apparently, however, no compelling public interest convinced them they could make juries believe that the defendants had acted unreasonably or that their religious freedom needed to be curtailed in the name of protecting some competing right.[44]

Curiously enough, by the mid-1960s COPs had even successfully persuaded the federal government and most states to include Christian Scientists in workers' sick-leave privileges, state workers' compensation laws, and government health care benefits, and many American insurance companies covered treatment by Christian Science practitioners.[45] Understandably confident that medical practice laws released religious healers from licensing requirements and that parental and religious rights protected believing parents, Christian Scientists slid into a false sense of security about their legal status.[46]

A stunned Church of Christ, Scientist, awakened from its slumber and leaped into action, however, after a Massachusetts court convicted Christian Scientist Dorothy Sheridan in 1967 of manslaughter for failing to treat her five-and-a-half-year-old daughter for pneumonia. Rather than pursuing an appeal that might overturn Sheridan's conviction on a technicality but affirm the parental obligation to seek medical care, the COP moved to change the law. It took time and considerable behind-the-scenes pressure, but in 1971 the governor of Massachusetts signed a law stating: "A child shall not be deemed to be neglected or lack proper physical care for the sole reason that he is being provided remedial treatment by spiritual means alone in accordance with the tenets and practices of a recognized church or religious denomination by a duly-accredited practitioner thereof."[47] Soon forty-four states adopted similar laws that exempted parents from providing medical treatment when they furnished care according to their religious beliefs, and it seemed to be safe for Christian Scientists to continue their healing practices.[48]

Clearly Sheridan's conviction stunned Christian Scientists, but in hindsight they could have been better prepared. Developments before 1967 in the way courts struggled to balance three competing legal principles—parental authority over children, the state's power to protect the health and well-being of its citizens, and the rights guaranteed under the First Amendment—and the way the United States Supreme Court determined that balance in 1944 should have given Scientists pause.

Between 1930 and 1960 courts had decided in several successive cases involving religious healing that when a child was ill, the *parens patriae* doctrine should not just be used to punish "the parent after the child is dead" but should be used to "take swift steps to keep the child alive."[49] And Judge Wiley B. Rutledge, writing in *Prince v. Massachusetts* (1944) for the United States Supreme Court, affirmed *parens patriae* by stating "that the state has a wide range of power for limiting parental freedom and authority in things affecting the child's welfare" and affirmed *Reynolds v. United States* (1878) by explicitly stating "that this includes, to some extent, matters of conscience and religious conviction." Rutledge also directly linked health and religion by asserting that "the right to practice religion freely does not include liberty to expose the community or the child to communicable disease or the latter to ill health or death." Rutledge concluded: "Parents may be free to become martyrs themselves. But it does not follow they are free, in identical circumstances, to make martyrs of their children before they have reached the age of full and legal discretion when they can make that choice for themselves."[50]

The Sheridan case made it clear that even the retreat by Christian Scientists from the school of medicine to the sanctuary of religion had not guaranteed lives without trial. But a subsequent decision by the United States Supreme Court held renewed if somewhat ambiguous promise. In *Wisconsin v. Yoder* (1972) the court rejected "the older rationale that the state is free to regulate religiously-motivated actions," but it affirmed the states' right to judge such actions in the light of their "undoubted power to promote health, safety and general welfare."[51] Therefore if the state could demonstrate that Christian Science healing had not worked, and as a result had been detrimental to a child's health, it could override the right to practice religious healing and move to convict parents who had practiced it on their children. A clear onus now rested on Christian Scientists to demonstrate to the state's satisfaction that Christian Science healing worked. But given their refusal to submit Christian Science healing to double-blind studies, they faced a seemingly impossible challenge.

Child Abuse, Medical Care, and the Law

Then, in 1979 a Phil Donahue television show featured Douglas and Rita Swan, two angry Christian Science parents who, despite practitioner treatment, had lost their fifteen-month-old son to meningitis. That show, according to at least one influential Christian Scientist, ignited a "nation-wide attack on Christian Science" and ushered in a decade of legal persecution unseen since the early 1900s.[52] That the show alone unleashed a new national assault on Christian Scientists seems doubtful, but the 1980s did reintroduce Christian Scientists to a life on trial that many had forgotten. And in the particulars of the Swan family experience reappeared a pattern of events much like that of a century before.

In June 1977 the Swans' infant son, Matthew, suffering from an intense and recurring fever, received Christian Science treatment.[53] Given Matthew's infancy and the crucial role his parents' thoughts would play in a successful demonstration over the error of belief, the practitioner affirmed in their presence: "Matthew, you can't be sick. God is your life. Your life can't be subject to the laws of matter. There is no death." Despite these words, however, Matthew's mother, a lifelong Scientist, expressed her fears that Christian Science was not "getting to the root cause of the problem." Practitioners had treated her son for the fever on three previous occasions, but it kept recurring. Believing that the fears of the parents were the real source of Matthew's apparent difficulties, the practitioner asked: "What could medicine do for Matthew? I suppose the doctors would give him a baby aspirin. You know that wouldn't solve the real problem."

Days passed with the parents agonizing over Matthew, practitioners doing their work, and Matthew's condition worsening. With their son screaming in pain one night, the parents almost turned to a physician, but their current practitioner convinced them that Matthew was being healed and that they only needed the ministrations of a Christian Science nurse. When the nurse arrived, she soothed and comforted the Swans and recommended orange juice for Matthew; after she had gone, Matthew's pain appeared to have subsided, "even though he was delirious and totally incoherent."

Several days later the Swans' current practitioner paid them a house call, looked at Matthew, and announced: "He used to be lethargic and paralyzed. Now look at him, he can move so freely! Everything looks lovely now . . . except the head." Rita, however, saw a different Matthew, "totally incoherent, thrashing around wildly."

As Matthew's condition deteriorated further, the practitioner focused on his listless head and speculated that a fall may have broken a bone in his neck. She recommended that the parents have his neck X-rayed. Their visit to a hospital emergency room for the X rays alerted medical staff to his condition, but despite their best efforts Matthew died of meningitis five days later, three weeks after the onset of his fever. Neither Christian Science nor medical science could save Matthew.

Charging "negligence and misrepresentation," the Swans first sued the Mother Church and the practitioners who had unsuccessfully treated their child, then they founded CHILD (Children's Healthcare Is a Legal Duty, Inc.) to mount a nationwide crusade against all religious healers who ignored medical science.[54] By 1989 the organization had identified 137 children (seven Christian Scientists) over the past sixteen years who had died after failing to receive medical treatment for religious reasons. Armed with that knowledge and joined by the American Medical Association, the American Academy of Pediatrics, the National District Attorneys Association, and child advocacy groups, CHILD successfully prodded several states to revoke statutes that exempted parents from child abuse laws for religious reasons.[55]

The efforts of CHILD and others to publicize the failures of Christian Science healers and to reform medical laws created a public relations challenge for the church, but nothing like the financial and legal pressures forced on the movement during the 1980s, when officials charged Christian Science parents with felonies in seven cases across the country (see appendix). Part of a nationwide trend, between 1982 and 1990 American prosecutors filed twenty-nine felony charges against religious healers in the death of children and won nineteen convictions.[56]

Shauntay, the four-year-old daughter of Laurie G. Walker, died of bacterial meningitis more than two weeks after she first exhibited flulike symptoms, despite Christian Science treatments.[57] Charged with second-degree murder (later dismissed), involuntary manslaughter, and child abuse for actions that Christian Scientists and many others believed the law permitted, Laurie Walker came to represent for many the injustice that can come when the state tramples on religious freedom and due process.[58]

At issue in the Walker case, as in two other 1984 California cases against Scientists, was the intent of the 1976 amendment to section 270 of the California penal code, passed with the advice of Christian Science lobbyists, that stated that parents who treated their sick children "by spiritual means through prayer alone" fulfilled their duty to provide "remedial care" and were thus exempt from prosecution for misdemeanor child abuse.[59] Confident that the statute made it clear that Walker had not broken the law by pursuing Christian Science healing, her defense team appealed to the state supreme court an unfavorable ruling by a district court of appeal. In a startling decision, state supreme court justice Stanley Mosk, writing for the court, stated: "Parents have no right to free exercise of religion at the price of a child's life, regardless of the prohibitive or compulsive nature of the governmental infringement." Moreover, although the law in question may have precluded prosecution for a misdemeanor, it did not proscribe the prosecution of a felony. Mosk concluded: "When a child's health is seriously jeopardized, the right of a parent to rely on prayer must yield."[60]

Initially too stunned to comment, officials of the Mother Church must have wondered if all the hard-fought legal battles, the ceaseless activities of the COPs, and the seemingly endless road of accommodation had been for naught. After some reflection, however, they decided that nothing remained but to forge ahead. Walker's attorneys appealed to the United States Supreme Court, but they received another setback when the Court declined to consider the case. Although the Court rarely explains its reasons for not accepting a case, legal experts took the decision to mean that it believed the Constitution "protects religious beliefs but not religious conduct." Apparently the justices decided in this case to side with *Yoder*'s affirmation of the states' "power to promote health." With all constitutional grounds for appeal exhausted, Walker returned to Judge George Nicholson's Sacramento County superior courtroom for trial. In June 1990 the court found her guilty of involuntary manslaughter, placed her on probation, and sentenced her to six hundred hours of community service.

Like the Walker case, the other cases against Christian Scientists in the 1980s bore many resemblances to those of the century's first decade. Each involved the death of a child; each child died of a disease that the medical community believed it could have successfully treated; and each involved criminal charges against the dead child's parents (see appendix). But differences appeared as well. Charges still included murder and manslaughter, but usually the state now singled out child abuse as the key offense, and in no case did it bring charges of illegal medical practice. And rather than spearheading these actions against Christian Scientists, physicians joined the charge well after the first battles had been joined.[61]

A 1962 article in *JAMA* graphically described "the battered child syndrome" in the United States and documented the widespread failure of medical personnel to report cases to authorities.[62] Subsequent studies confirmed the article's conclusions and together with media support stimulated state and federal governments to establish guidelines for the "mandatory physician reporting of suspected cases," which forty-seven states had implemented by 1965. The new legislation concerned COPs, who mounted campaigns in their respective states to amend the laws to ensure that treatment of children by Christian Science parents or practitioners did not constitute child abuse or neglect. The COPs found little success, however, until they persuaded Congress to mandate such exemptions in the regulations implementing the 1974 Child Abuse Prevention and Treatment Act. Forty-three states had passed the exemptions by 1983.

In that same year the shocking McMartin Preschool child abuse case in California, the first of several nationwide, added to the growing American awareness—some said paranoia—about child abuse. Irrational urges to hurt children apparently lurked inside caregivers and parents, and only an ever vigilant public could ensure the safety of its children. Such widely publicized and often sensationalized events created a climate that only made it more difficult for the little understood worldview of Scientist parents to receive a sympathetic hearing from prosecutors, judges, or juries.[63]

To be sure, along with their colleagues' resistance to all forms of irregular medicine, individual physicians had continued since the 1920s to single out religious healing for opposition, but organized medicine had launched no national campaigns against Christian Science. Similarly, in the early 1980s organized medicine lingered in the background, with only individual physicians serving as expert witnesses once the trials began. That all changed in the midst of the criminal proceedings, however, when the American Academy of Pediatrics, criticized earlier by some for having moved too late to recognize and prevent child abuse, released its 1988 recommendation that Americans should not allow religious beliefs as a justification for child abuse or neglect brought about by refusing medical care.[64] In the words of Norman C. Fost, University of Wisconsin pediatrician and ethicist and chief architect of the academy's recommendation, "Are there not instances where we hold adults liable for failing to do what average, informed persons would do, and isn't one of those times when there's a need to seek the best medical help available to preserve the health of the young? . . . Are we to permit children to become martyrs to their parents' faith?"[65] By 1994 three states had changed laws that had previously "allowed people to deny their children medical care for religious reasons."[66]

Christian Scientists, of course, viewed things differently. Complaining that they were being unfairly judged based on a handful of tragedies, they cited statistics to demonstrate that Christian Science healing provided the best way to "preserve the health of the young." Doubtless many Scientists agreed with church leader Nathan A. Talbot that the efforts of pediatricians and others were "nothing more than a desire to prosecute out of existence a type of healing that does not fit into the scientific worldview and scheme of things." True, pundits and many members of the legal community called for the abolition of religious healing, but others simply suggested that Christian Scientists pursue their practices "in conjunction with regular medical attention," thereby emulating many New Thought and New Age groups but also continuing down the accommodationist path earlier taken by Mrs. Eddy, Judge Clifford P. Smith, and others in Christian Science.[67] In fact Christian Scientists had pursued such a course in Great Britain and Canada since the affirmation of *Rex v. Lewis* in 1903. But accommodation may be able to go only so far before it becomes appeasement or, worse, complete surrender. An aging but still influential Robert Peel cautioned fellow believers against mixing material and spiritual healing, preferring to believe that Scientists had brought the persecution on themselves precisely because many had given up a radical dependence on spiritual healing to resolve conflict and demonstrate true healing.[68]

Whatever the reasons for the upsurge in Christian Science prosecutions, Scientists had been wounded by the publicity surrounding the criminal trials, and in response they flooded the press with justifications for their actions and evangelized for Christian Science healing in the modern world.[69] Possibly owing in part to such efforts but also because of the chilling effects the criminal conviction of parents had on Christian Science practices, the flurry of Christian Science trials in the 1980s subsided in the early 1990s, and concerns about a Christian Science threat to public health or children's safety virtually disappeared from the public and medical press by 2000.[70] But the church had suffered severe damage to more than its public reputation. An enumeration of the practitioners listed in the January 1985 *Journal* revealed that in the United States they numbered 2,884, of whom 32 percent practiced in the four western states of California (725), Washington (82), Oregon (63), and Arizona (61). But in fifteen years the numbers had tumbled to just 1,396 practitioners in the United States (31 percent in the same four western states)—an overall decline of 52 percent.[71]

To make matters worse, a new set of challenges to the survival of Christian Science arose from within the church. Unknown to many of the church's members, while the church had been fighting its legal battles, the Mother Church's operations had been drowning in red ink, and rumors of financial malfeasance by church administrators spread quickly. Not only did severe cutbacks in operating funds dramatically alter the way the church carried out its mission, but church leaders settled on what for many church members appeared to be a Faustian agreement. In order for the church to receive a multimillion-dollar bequest, the board of directors agreed to distribute officially a controversial biography of Mary Baker Eddy that the church had judged heretical forty-five years earlier for declaring Eddy the equal of Jesus.[72] The board must have thought its action would alleviate the church's financial distress, but in the words of Christian Science scholar Stephen Gottschalk, the decision to release the biography showed that the board had "chosen to deliberately flout" the *Church Manual* and became "central to the constitutional crisis facing the church."[73]

The public airing of such church family squabbles and the sensational trials of Christian Science parents seemed strangely out of character for a movement whose members, during the past eighty years, had grown to blend into the regular patterns of American life. But for those familiar with the tumultuous disputes of the 1880s and 1910s, conflicts over church authority and confrontations with public officials had been an almost routine part of church life. Ironically, although Christian Scientists had grown to represent for many Americans the most urbane and sedate example of the often peculiar yet perennial practice of religious healing, their distinctive practices of physical healing continued to bring them their greatest public notoriety and present one of the greatest challenges for their continued survival as a movement.

Appendix

Court Cases Involving Christian Science Practice

Date	Name	Charge	Child?	Diagnosis	Verdict	Location
1887[1]	Charlotte Eddy Post (practitioner)	Practicing medicine without a license	No	Unknown	Guilty (1887); overturned (1887)	McGregor, Iowa
1887[2]	Fred W. Bunnell (practitioner)	Practicing medicine and obstetrics without a license	Yes, in some cases	Some cases involved childbirth	Not guilty (1887)	Kearney, Nebraska
1887[3]	Emma Whitlock (practitioner)	Practicing medicine without a license	No	Unknown	Not guilty (1887)	Fredericton, New Brunswick, Canada
1888[4]	George W. Wheeler (practitioner)	Right to compensation	No	—	Found for plaintiff (1888)	Wilton, Maine
1888[5]	Charles M. Howe (practitioner)	Unknown	Unknown	Unknown	Found for defendant (1888)	St. Joseph, Missouri
1888[6]	Abby H. Corner (practitioner)	Manslaughter	Yes	Childbirth	Not guilty (1888)	West Medford, Massachusetts
1889[7]	John H. and Isabella M. Stewart (practitioners)	Practicing medicine without a license	Yes	Unknown (several)	Guilty; overturned	Toronto, Ontario, Canada
1891[8]	Eliza Ward (practitioner)	Manslaughter	No	Meningitis	Not guilty (1892)	Riverside, California
1893[9]	Ezra M. Buswell (practitioner)	Practicing medicine without a license	Yes	Cholera infantum, diptheria, snake-bite, open sore, rheumatism	Not guilty (1893); reversed (1894)	Beatrice, Nebraska
1895[10]	Hattie E. Graybill (practitioner)	Practicing medicine without a license	Unknown	Unknown	Not guilty (1895)	Atchison, Kansas
1895[11]	John R. Hatten (practitioner)	Manslaughter	Yes	Tubercular meningitis	Not guilty (1895)	Dayton, Ohio

Year	Name	Charge	Patient died	Cause of death/illness	Verdict	Location
1895[12]	Charles A. Owen (practitioner)	Failure to report diphtheria	Yes	Malignant diphtheria	Guilty (1895); reversed (1895)	Davenport, Iowa
1895[13]	Mary Ellen Beer (practitioner)	Manslaughter	Yes	Nonmalignant diphtheria	Not guilty (1895)	Toronto, Ontario, Canada
1895[14]	Laura B. Aikin (practitioner)	Manslaughter	No	Childbirth; peritonitis	Not guilty (1895); before trial prosecutor recommended this verdict	Memphis, Tennessee
1895[15]	Jesse Samis (husband) and Richard A. Cook (practitioner)	Manslaughter	Yes	Childbirth; septicemia	Not guilty (1895)	Los Angeles, California
1896[16]	Agnes Chester (practitioner)	Practicing medicine without a license	Yes	Childbirth	Not guilty (?)	Kalamazoo, Michigan
1897[17]	Walter E. Mylod (practitioner)	Practicing medicine without a license	No	"Made up"*	Not guilty (1898)	Providence, Rhode Island
1897[18]	David Anthony (practitioner)	Practicing medicine without a license	No	"Made up"*	Not guilty (1898)	Providence, Rhode Island
1897[19]	Amanda J. Baird (practitioner)	Failure to report contagious disease	Yes	Malignant diphtheria	Guilty (1897); affirmed (1898); reversed (1902)	Kansas City, Missouri
1897[20]	Bertha H. Sessford (practitioner)	Practicing medicine without a license	Yes	Diphtheria	Not guilty (1897)	Washington, D.C.
1898[21]	Harriet O. Evans (practitioner)	Practicing medicine without a license	No	Typhoid fever	Guilty (1898); reversed 1899	Cincinnati, Ohio
1898[22]	Allie Putnam (practitioner)	Practicing medicine without a license	No	"Made up"*	Guilty (1898); reversed (1899?)	Cincinnati, Ohio
1899[23]	Benjamin J. Hammett (practitioner)	Practicing medicine without a license	No	Paralysis	Charges dismissed (1899)	Pawnee City, Nebraska

Date	Name	Charge	Child?	Diagnosis	Verdict	Location
1899[24]	Mary Brookins and Albert P. Meyer (practitioners)	Practicing medicine without a license	Yes	Malignant diphtheria	Demurrers granted (1899)	Minneapolis, Minnesota
1900[25]	Crecentia Arries and Emma Nichols (practitioners)	Practicing medicine without a license	Yes	Diphtheria	Guilty (1900); reversed (1901)	Milwaukee, Wisconsin
1901[26]	Charles D. Pease (practitioner)	Failure to report contagious disease	No	Consumption	Not guilty (1901)	New York City, New York
1901[27]	Irving C. Tomlinson (practitioner)	Negligent and fraudulent practice of Christian Science healing	No	Appendicitis	Found for defendant (1901); affirmed (1904)	Concord, New Hampshire
1901[28]	James Henry Lewis (parent)	Manslaughter	Yes	Diphtheria	Guilty (1901); affirmed (1903)	Toronto, Ontario, Canada
1902[29]	John and Georgiana Quimby (parents) and John Carroll Lathrop (practitioner)	Manslaughter in the second degree	Yes	Diphtheria	Demurrers sustained (1905)	White Plains, New York
1902[30]	Merrill and Clara Reed (parents)	Willful child neglect (failure to provide nursing and medical attention)	Yes	Diphtheria	Not guilty (1902)	Los Angeles, California
1903[31]	Oliver W. Marble (practitioner)	Practicing medicine without a license	Yes; no	Typhoid fever; rheumatism	Guilty (1903); reversed (1904); exceptions sustained (1905)	Sandusky, Ohio

1907[32]	Edwin W. and Mary Watson (parents)	Manslaughter	Yes	Cerebrospinal meningitis	Guilty (1907); reversed (1909)	Mount Holly, New Jersey
1907[33]	Clarence W. Byrne (parent)	Misdemeanor failure to provide medical attendance for a minor	Yes	Bronchial pneumonia	Guilty (1907)	New York City, New York
1909[34]	Julius Benjamin (practitioner)	Malpractice	No	Sore on foot that led to amputation	Found for plaintiff (19??); affirmed (19??); reversed and new trial granted (1909)	New York City, New York
1911[35]	Willis V. Cole (practitioner)	Practicing medicine without a license	Yes	"Made up"*	Hung jury (1911); guilty (1912); affirmed (1914); reversed (1916)	New York City, New York
1920[36]	Mr. and Mrs. Andrew Walker (parents)	Manslaughter	Yes	Diphtheria	Husband guilty and fined $1,000; wife not guilty (1920); appeal filed, results ??	Newark, New Jersey
1925[37]	Mr. and Mrs. Robert Watson (parents)	Manslaughter	Yes	Diphtheria	Guilty; reversed (1925)	Manitoba, Canada
1967[38]	Dorothy Sheridan (parent)	Manslaughter	Yes	Pneumonia	Guilty and sentenced to five years' probation (1967)	Harwich, Massachusetts

Appendix (Continued)

Date	Name	Charge	Child ?	Diagnosis	Verdict	Location
1984[39]	Laurie G. Walker (parent)	Second-degree murder (later dismissed), involuntary manslaughter, and child abuse	Yes	Bacterial meningitis	Guilty (1990)	Sacramento County, California
1984[40]	Eliot and Lise Glaser (parents)	Second-degree murder, an alternative count of involuntary manslaughter, felony child endangering and failure to provide the necessities of life for their son	Yes	Bacterial meningitis	Acquitted (1990)	Los Angeles, County, California
1984[41]	Mark Rippberger and Susan Middleton (parents)	Involuntary manslaughter and child abuse	Yes	Bacterial meningitis	Guilty of child endangerment; cleared of involuntary manslaughter (1989)	Sonoma County, California
1984[42]	Ted B., et al. (parents)	Child placed in protective custody	Yes	Retinal blastoma	Declared dependent of the state; affirmed (1985); affirmed by state court of appeals (1987)	Contra Costa County, California

Year	Name	Charge	"Made up"*	Condition	Legal outcome	Location
1988[43]	John and Catherine King (parents)	Felony child abuse	Yes	Bone cancer	Pled "no contest" (1989)	Paradise Valley, Arizona
1988[44]	David and Ginger Twitchell (parents)	Manslaughter	Yes	Bowel obstruction	Guilty (1990); overturned by state supreme court (1993)	Boston, Massachusetts
1989[45]	William and Christine Hermanson (parents)	Felony child abuse and third-degree murder	Yes	Juvenile diabetes	Guilty (1987); affirmed (1990); overturned by state supreme court (1992)	Sarasota County, Florida
1989[46]	Mrs. McKown (mother) and stepfather	Second-degree manslaughter	Yes	Juvenile diabetes	Indictments overturned by the state supreme court (1991)	Minnesota
1990[47]	Kara and Morris Newmark	Suit for custody to treat	Yes	Lymphoma	Custody awarded to state (1990), immediately stayed to permit appeal; reversed by state supreme court (1990)	Delaware

*"Made up" means that an agent of the state "made up" symptoms for the practitioner to treat.

1. *Christian Science Journal* 5 (1887): 161, 213, 269.

2. *Christian Science Journal* 5 (1887): 336–38.

3. *Christian Science Journal* 5 (1887): 166–72, 377.

4. Case file, "George W. Wheeler v. Prince A. Sawyer," March Term, 1888, Supreme Judicial Court, Franklin County, Maine. Maine State Archives, Augusta, Maine; Wheeler v. Sawyer, 15 Atl. (Me.) 67 (1888).

5. *Christian Science Journal* 6 (1888): 97.

Appendix (Continued)

6. *Boston Daily Globe*, 21 April 1888, evening edition, 1; 22 April 1888, morning edition, 2; 23 April 1888, evening edition, 1; 24 April 1888, morning edition, 2; 24 April 1888, evening edition, 5; 25 April 1888, morning edition, 4; 30 April 1888, morning edition, 3; 19 May 1888, morning edition, 4; 22 May 1888, evening edition, 5; 26 May 1888, evening edition, 8; *Christian Science Journal* 6 (1888): 258.

7. *Christian Science Journal* 6 (1889): 589; 7 (1889): 103–4, 226; 16 (1898): 92–95.

8. People of the State of California v. Eliza Ward, Superior Court, County of San Bernardino, SC 4230, 1891, Master Criminal Index. San Bernardino County Courthouse, San Bernardino, California; *San Bernardino (Calif.) Daily Courier*, 3 February 1892, 2; 3 February 1892, 3; 4 February 1892, 3; 5 February 1892, 3; 6 February 1892, 3; 7 February 1892, 2; *Christian Science Journal* 11 (1893): 14–17.

9. Transcript of Proceedings, State of Nebraska v. E. M. Buswell, District Court for Gage County. Nebraska State Historical Society, Lincoln, Nebraska, RG 69, Nebraska Supreme Court, case 6495 (box 20); State v. Buswell, 40 Neb. 158, 58 NW 728 (1894), 24 L.R.A. 68; *Beatrice (Nebr.) Daily Express*, 28 February 1893, 1, 4; 1 March 1893, 1; 2 March 1893, 1; 3 March 1893, 1; 4 March 1893, 4; 6 March 1893, 4; *Beatrice (Nebr.) Daily Times*, 28 February 1893, 4; 1 March 1893, 4; 2 March 1893, 4; 3 March 1893, 4; *Christian Science Journal* 11 (1893): 65–86, 91–94, 165–70; 12 (1894): 168–75.

10. *Atchison (Kans.) Daily Globe*, 16 May 1895, 2; *Christian Science Journal* 13 (1895): 164.

11. *Dayton (Ohio) Daily Journal*, 20 March 1895, 1, 4, 5; 21 March 1895, 1; 23 March 1895, 1; 25 March 1895, 4; 27 March 1895, 1; 28 March 1895, 1; 4 April 1895, 1; 8 April 1895, 1; 9 April 1895, 1; *Dayton (Ohio) Evening Herald*, 21 March 1895, 4; *Christian Science Journal* 13 (1895): 163–64.

12. Case file, City of Davenport v. C. A. Owen, case 1548. Clerk's Office, District Court, Scott County, Davenport, Iowa; *Davenport (Iowa) Democrat*, 2 June 1895, 1; 3 June 1895, 1; 5 June 1895, 2; 6 June 1895, 1; 9 June 1895, 1; 4 December 1895, 1; *Christian Science Journal* 13 (1896): 428.

13. Regina v. Beer, 32 Canada Law. J. 416 (1895); *Toronto World*, 30 October 1895, 1; 31 October 1895, 1; 1 November 1895, 2; 4 November 1895, 6; 5 November 1895, 4; 13 November 1895, 1; 6 December 1895, 2; 8 December 1895, 4; *Christian Science Journal* 13 (1896): 469–72.

14. *Christian Science Journal* 13 (1896): 475–76.

15. *Los Angeles Times*, 26 March 1895, 1, 5; 27 March 1895, 1, 9; 25 April 1895, 10.

16. *Kalamazoo (Mich.) Daily Telegraph*, 11 February 1896, 1; 12 February 1896, 1, 8; 21 February 1896, 1, 5; 3 March 1896, 1, 2; *Christian Science Journal* 14 (1896): 34–35.

17. Case file, State of Rhode Island, Gardner T. Swarts, Complainant, v. Walter E. Mylod, Respondent, case 18426. District Court of the Sixth Judicial District of the State of Rhode Island and Providence Plantations. Phillips Memorial Library Archives, Providence College, Providence, Rhode Island; State v. Mylod, 20 R.I. 632, 40 Atl. 753, 41 L.R.A. 428 (1898); *Providence (R.I.) Evening Bulletin*, 19 July 1898, 6; *Christian Science Journal* 16 (1898): 405–16, 447–49; *Christian Science Weekly*, 24 November 1898, 4–6; 1 December 1898, 5–6.

18. Case file, State of Rhode Island, Gardner T. Swarts, Complainant, v. David Anthony, Respondent, case 18425. District Court of the Sixth Judicial District of the State of Rhode Island and Providence Plantations. Phillips Memorial Library Archives, Providence College, Providence, Rhode Island; State v. David Anthony, per curiam Mylod, 20 R.I. 632, 40 R.I. 692, 40 Atl. 753, 41 L.R.A. 428 (1898); *Providence (R.I.) Evening Bulletin*, 19 July 1898, 6.

19. Case file, Kansas City v. Mrs. A. J. Baird, case 6082, box 12333. Missouri Record Management and Archives Service, Kansas City Court of Appeals, Transcript and Case Records, Jefferson City, Missouri; Kansas City v. Baird, 92 Mo. A. 204 (1902); *Kansas City (Mo.) Star*, 4 November 1897, 1; 5 November 1897, 1, 4; 6 November 1897, 2; 7 November 1897, 1; 8 November 1897, 1; 9 November 1897, 1; 10 November 1897, 1; 11 November 1897, 1, 4; 20 January 1898, 1.

20. *Washington, DC., Evening Star*, 15 December 1897, 2; *Christian Science Journal* 16 (1898): 143–48, 187–88.

21. Evans v. State, 9 Ohio Decisions 222 (1889 [*sic*]), 6 Ohio N.P. 129; *Cincinnati Enquirer*, 31 October 1898, 5; 7 November 1898, 5; 12 November 1898, 8; 14 November 1898, 6; 15 November 1898, 5; 16 November 1898, 12; 18 November 1898, 16; 21 November 1898, 6; 9 December 1898, 10; 31 December 1898, 2; *New York Times*, 12 November 1898, 6; 16 November 1898, 1; 24 December 1898, 6; *Christian Science Journal* 16 (1899): 827–35; *Christian Science Weekly*, 29 December 1898, 2–3.

22. *Cincinnati Enquirer*, 22 December 1898, 12; *New York Times*, 24 December 1898, 6, c. 4.

23. *Christian Science Journal* 18 (1900–1901): 16–20.

24. Case file, State of Minnesota v. Mary Brookins (indicted as Mary Brookings), case 6123, District Court for the Fourth Judicial District, County of Hennepin, Minnesota. Clerk's Office, District Court for the Fourth Judicial District, Minneapolis, Minnesota; case file, State of Minnesota v. Albert P. Meyer, case 6124, District Court for the Fourth Judicial District, County of Hennepin, Minnesota. Clerk's Office, District Court for the Fourth Judicial District, Minneapolis, Minnesota; *Minneapolis Journal*, 16 October 1899, 6; 25 October 1899, 1; 27 October 1899, 6; 9 November 1899, 6; 21 November 1899, 4; 25 November 1899, sec. 2, p. 8; *Christian Science Journal* 17 (1900): 696–701.

25. A. C. Umbreit, ed., *Report of the Trial of Creentia Arries and Emma Nichols, Charged with Practicing Medicine without a License, Before the Police Court of the City of Milwaukee, Giving a Verbatim Report of All the Testimony, Rulings of the Court, Arguments of Counsel and Decisions of the Presiding Judge* (Milwaukee: Edw. Keogh Press, 1900); *Milwaukee Journal*, 10 May 1900, 1; 11 May 1900, 1, 9; 12 May 1900, 5; 2 June 1900, 1, 3; 4 April 1901, 1; 15 April 1901, 5; *Christian Science Sentinel*, 2 May 1901, 556–58.

26. *New York Times*, 20 December 1900, 6; 22 December 1900, 6; 23 December 1900, 18; 31 January 1901, 3; 6 February 1901, 6; 12 February 1901, 16; 14 February 1901, 6; 1 March 1901, 2.

27. Jennie A. Spead v. Irving C. Tomlinson, *State of New Hampshire Supreme Court Briefs and Cases*, 226:283–503. Supreme Court Library, Concord, New Hampshire; Spead v. Tomlinson, 73 N.H. 46, 59 Atl. 376 (1904); 68 L.R.A. 432; *Christian Science Journal* 22 (1904): 529–31.

28. Rex v. Lewis, 6 Ont. L. R. 132, 1 B.R.C. 732 (1903), 7 Can. Crim. Cases 261, Anno. 10 ALR 1150; *Toronto World*, 2 November 1901, 9; 4 November 1901, 6; 5 November 1901, 5; 6 November 1901, 7; 7 November 1901, 4; 30 June 1903, 4; *Christian Science Sentinel*, 19 December 1901, 247–57; 9 January 1902, 301–2.

29. *New York Evening Journal*, 22 October 1902, 4; 23 October 1902, 8; 24 October 1902, 5; 25 October 1902, 4; 27 October 1902, 3, 12; 28 October 1902, 5; 30 October 1902, 8; 1 November 1902, 11; 6 November 1902, 6; 25 November 1902, 11; 26 November 1902, 5; *New York Times*, 22 October 1902, 16; 23 October 1902, 16; 24 October 1902, 1; 25 October 1902, 6; 27 October 1902, 14; 29 October 1902, 2; 31 October 1902, 1; 26 November 1902, 1; 8 August 1905, 7; *Christian Science Sentinel*, 12 August 1905, 804.

30. *Los Angeles Daily Times*, 26 November 1902, sec. 2, p. 2; 27 November 1902, sec. 2, p. 2; 29 November 1902, sec. 2, p. 2; 30 November 1902, sec. 4, p. 8; *Christian Science Journal* 20 (1903): 669–83, 684–705; *Christian Science Sentinel*, 18 December 1902, 248.

31. Case file, State of Ohio v. Oliver W. Marble, no. 9002, January Term, 1904, Supreme Court of the State of Ohio, Columbus, Ohio; State v. Marble, 72 Ohio St. 21, 73 N.E. 1063, 70 L.R.A. 835, 106 Am. St. Rep. 570 (1905); *Sandusky (Ohio) Daily Register*, 14 June 1903, 1; 15 June 1903, 1; 14 July 1903, 2; 12 August 1903, 8; 13 August 1903, 5; 14 August 1903, 3; 15 August 1903, 3; 18 August 1903, 3; 20 August 1903, 3; 1 March 1905, 3.

32. *New York Times*, 28 May 1907, 1; 1 June 1907, 1; 10 October 1907, 18; 11 October 1907, 1; 13 October 1907, 1; 18 October 1907, sec. 3, p. 8; 24 October 1907, 1; 6 November 1908, 14; 24 February 1909, 5.

33. *New York Times*, 18 May 1907, 1; 28 June 1907, 3; 29 June 1907, 16; 3 August 1907, 14; 5 August 1907, 6.

34. Raisler v. Benjamin, 118 N.Y.S. 223 (1909), 133 App. Div. 721.

35. Transcript of Proceedings, People v. Willis V. Cole. Transcript 1767, March 28, 1912 [*sic*], Willis V. Cole, Judge Seabury, Collation 227. Library of the John Jay College of Criminal Justice, New York, N.Y.; People v. Cole, 148 N.Y.S. 708, 163 App. Div. 292, 31 N.Y. Cr. 487 (1911, 1912, 1914), reversed 219 N.Y. 98, 113 N.E. 790, L.R.A. 1917C 816 (1916), 25 N.Y. Cr. R. 350; *New York Times*, 15 January 1911, 1; 16 January 1911, 10; 27 January 1911, 6; 28 January 1911, 7; 29 January 1911, 5; 12 February 1911, 1; 22 March 1911, 2; 12 April 1911, 9; 6 May 1911, 6; 20 June 1911, 5; 1 March 1912, 7; 2 March 1912, 9; 31 March 1912, 9; 11 July 1914, 4; 13 July 1914, 8; 4 October 1916, 10; 10 October 1916, 11; 11 October 1916, 10.

36. *New York Times*, 14 May 1919, 8; 5 May 1920, 11; 6 May 1920, 1, 10; 11 May 1920, 1.

37. Rex v. Elder, 35 Manitoba 161, 44 Can. Cr. C. 75 (1925).

38. Leo Damore, *The* "CRIME" *of* Dorothy Sheridan (New York: Arbor House, 1978).

39. Walker v. Superior Court of Sacramento County, 47 Cal. 3d 112, 253 Cal. Rptr. 1, 763 P2d 852 (1988), *cert den.* 105 LE2d 695, 109 S.Ct. 3196 (1989); People v. Walker, Cal. Superior Ct. (21 July 1990); *Riverside County (Calif.) Press-Enterprise*, 28 March 1986, A-4; 26 August 1986, A-8; *Los Angeles Times*, 9 March 1988, sec. 1, p. 23; 11 November 1988, sec. 1, pp. 1, 24; 20 June 1989, sec. 1, p. 16; 25 June 1990, A20.

40. State v. Glaser, no. A–753942 (Cal. Super. Ct., 16 February 1990); *Orange County (Calif.) Register*, 23 June 1984; *Riverside County (Calif.) Press-Enterprise*, 14 March 1985, A-8; 26 August 1986, A-8; *Los Angeles Times*, 18 February 1990, sec. B, pp. 1, 5; Richard J. Brenneman, "Nestling's Faltering Flight: The Short Life and Death of Seth Ian Glaser," in *Deadly Blessings: Faith Healing on Trial* (Buffalo, N.Y.: Prometheus Books, 1990).

41. State v. Rippberger, no. 19301-C (Cal. Super. Ct., 2 November 1989); *Riverside County (Calif.) Press-Enterprise*, 28 March 1986, A-4; 26 August 1986, A-8; 5 August 1989, A-3; *Los Angeles Times*, 27 June 1989, sec. 1, pp. 1, 16.

42. In re. Eric B., a Person Coming under the Juvenile Court Law, Contra Costa County Department of Social Services v. Ted. B., et al., no. A031383 (189 Cal. App. 3d 996; Cal. App. LEXIS 1426; 235 Cal. Rptr. 22). Appeal from Superior Court of Contra Costa County, no. JUV 69430.

43. State v. King, no. CR 88–07284 (Ariz. Sup. Ct., 1 September 1989); *Los Angeles Times*, 27 June 1989, sec. 1, pp. 1, 16.

44. Commonwealth v. Twitchell, no. 89–210 (Mass. Sup. Jud. Ct., 4 July 1990); *Insight*, 20 June 1988, 57; *Los Angeles Times*, 27 June 1989, sec. 1, pp. 1, 16; 5 July 1990, A26; 12 August 1993, A32.

45. State v. Hermanson, no. 86–3231 (Fla. 12th Cir. Ct., 1987); Hermanson v. State of Florida, 570 So. 2d 322 (Fla. App. 2 Dist., 1990).

46. State of Minnesota v. McKown, 461 N.W. 2d 720 (Minn. Ct. App., 1990); State of Minnesota v. McKown, no. 89052954 (Minn. District Ct., 2 April 1990); Minnesota v. McKown, WL 183119 (Sup. Ct. Minn., 29 September 1991); *Los Angeles Times*, 21 September 1991, A25.

47. Kara and Morris Newmark v. Teresa Williams/DCPS [Delaware Division of Child Protective Services], no. 325, 1990, Supreme Court of Delaware. 588 A. 2d 1108; 1991 Del. LEXIS 104; 21 A.L.R. 5th 857. Appeal from Williams v. Newmark, Delaware Family Court, no. CN90–9235, Conner, J, slip op. (12 September 1990).

Notes

Introduction

1. Claudia Wallis and Jeanne McDowell, "Faith and Healing," *Time*, 24 June 1996, 58. This was a Time/CNN poll of 1,004 Americans conducted by Yankelovich Partners.

2. "Center for The Study of Religion/Spirituality and Health," www.geri .duke.edu/religion/Homepage.htm, 9 October 2001; "mbmi.org, The Mind Body Medical Institute," www.mbmi.org, 9 October 2001; "The University of Arizona Program in Integrative Medicine," integrativemedicine.arizona.edu.

3. "Our Mission," www.nihr.org/default.asp, 9 October 2001.

4. "Clinicians and Clergy to Discuss Role for Both Medical and Spiritual Care," 7 March 2001, www.meta-list.org.

5. For some recent reflections on the nature of scientific authority in America, see Ronald G. Walters, ed., *Scientific Authority and Twentieth-Century America* (Baltimore: Johns Hopkins University Press, 1997).

6. For an overview of the models for understanding the relation between science and religion, see David B. Wilson, "The Historiography of Science and Religion," Colin A. Russell, "The Conflict of Science and Religion," and Stephen C. Meyer, "The Demarcation of Science and Religion," all in *The History of Science and Religion in the Western World: An Encyclopedia*, ed. Gary B. Ferngren, Edward J. Larson, and Darrel W. Amundsen (New York: Garland, 2000), 3–23.

7. By 1906 Christian Science membership soared from only about 1,500 orthodox Scientists in 1890 to about 40,000 (72 percent of them women). Regarding the calculation of these numbers, see chapter 2, notes 85 and 86.

8. For a detailed analysis of the rhetorical patterns of famous trials, see Janice Schuetz and Kathryn Holmes Snedaker, *Communication and Litigation: Case Studies of Famous Trials* (Carbondale: Southern Illinois University Press, 1988). For a helpful introduction to ways the social history of law and medicine have enlivened recent medicolegal history, see Michael Clark and Catherine Crawford, eds., *Legal Medicine in History* (New York: Cambridge University Press, 1994).

9. W. Lance Bennett and Martha S. Feldman, *Reconstructing Reality in the Courtroom: Justice and Judgment in American Culture* (New Brunswick, N.J.: Rutgers University Press, 1981), quotations on 164–65, 167.

10. For an outstanding study of the way medical authority, gender, and public opinion interacted in a late nineteenth-century courtroom, see Regina Morantz-Sanchez, *Conduct Unbecoming a Woman: Medicine on Trial in Turn-of-the-Century Brooklyn* (New York: Oxford University Press, 1999).

11. Michael R. Belknap, ed., *American Political Trials* (Westport, Conn.: Greenwood Press, 1981), 6. For a slightly different definition of political trial that gives more emphasis to the disparity of power between the contending parties, see Theodore L. Becker, ed., *Political Trials* (Indianapolis: Bobbs-Merrill, 1971), xi–xii.

12. Robert Hariman, ed., *Popular Trials: Rhetoric, Mass Media, and the Law* (Tuscaloosa: University of Alabama Press, 1990), 26. See also Frank Luther Mott, *American Journalism: A History of Newspapers in the United States through 250 Years, 1690 to 1940* (New York: Macmillan, 1947); John Lofton, *Justice and the Press* (Boston: Beacon Press, 1966); and Lawrence M. Bernabo and Celeste Michelle Condit, "Two Stories of the Scopes Trial: Legal and Journalistic Articulations of the Legitimacy of Science and Religion," in Hariman, *Popular Trials*, 55–85, 204–18.

13. Numerous sources document this late-century change. For example, see Ronald L. Numbers and Darrel W. Amundsen, eds., *Caring and Curing: Health and Medicine in the Western Religious Traditions* (New York: Macmillan, 1986; reprint, Baltimore: Johns Hopkins University Press, 1997); "Part IX. Medicine and Psychology," in Ferngren, Larson, and Amundsen, *History of Science and Religion in the Western Tradition*. For the debate over medicine and prayer, see Larry Dossey, *Healing Words: The Power of Prayer and the Practice of Medicine* (San Francisco: HarperCollins, 1993), and the wide-ranging responses it has received.

14. Robert Peel, *Christian Science: Its Encounter with American Culture* (New York: Henry Holt, 1958), 11.

15. Portions of my overview of Christian Science history in this introduction draw on my article "The Christian Science Tradition," in *Caring and Curing: Health and Medicine in the Western Religious Traditions*, ed. Ronald L. Numbers and Darrel W. Amundsen (New York: Macmillan, 1986; reprint, Baltimore: Johns Hopkins University Press, 1997), 421–46.

16. For a good overview of New Thought, see Beryl Satter, *Each Mind a Kingdom: American Women, Sexual Purity, and the New Thought Movement, 1875–1920* (Berkeley: University of California Press, 1999). See also Charles S. Braden, *Spirits in Rebellion: The Rise and Development of New Thought* (Dallas: Southern Methodist University Press, 1963), 89–128.

17. Mary Baker Eddy, *Church Manual of the First Church of Christ, Scientist, in Boston, Massachusetts*, 89th ed. (Boston: Trustees under the Will of Mary Baker G. Eddy, 1925), 15–16.

Chapter One. Mary Baker Eddy: Patient, Healer, Teacher

Epigraph: *Banner of Light* 23 (June 20, 1868): 8; *Banner of Light* 23 (June 27, 1868): 7; *Banner of Light* 23 (July 4, 1868): 7.

1. This may be why Choate rendered a final decision in the case only for Stanley and not for Tuttle.

2. The foregoing paragraphs are based on the following: Eddy v. Tuttle et al. and trs., case 181, Supreme Judicial Court, Salem, Mass., 1880–81, vol. DD, 115– 20, and George F. Choate's minutes of the case as quoted in Georgine Milmine, *The Life of Mary Baker G. Eddy and the History of Christian Science* (New York: Doubleday, Page, 1909; reprint, Grand Rapids: Baker Book House, 1971), 139–46 (page references are to reprint edition). Milmine's work has recently been re-issued as Willa Cather and Georgine Milmine, *The Life of Mary Baker G. Eddy and the History of Christian Science* (Lincoln: University of Nebraska Press, 1993).

3. For further discussion of the relationship between Eddy and Daniel H. Spofford, see Robert Peel, *Mary Baker Eddy* (New York: Holt, Rinehart, and Winston, 1966–77), 2:40, 42–44, 50–58, and Gillian Gill, *Mary Baker Eddy* (Reading, Mass.: Perseus Books, 1998), esp. chaps. 13–14, pp. 234–70.

4. The following biographical paragraphs on Mary's childhood and youth are patterned after Ronald L. Numbers and Rennie B. Schoepflin, "Ministries of Healing: Mary Baker Eddy, Ellen G. White, and the Religion of Health," in *Women and Health in America: Historical Readings*, 2d ed., ed. Judith Walzer Leavitt (Madison: University of Wisconsin Press, 1999), 579–95; Rennie B. Schoepflin, "The Christian Science Tradition," in *Caring and Curing: Health and Medicine in the Western Religious Traditions*, ed. Ronald L. Numbers and Darrel W. Amundsen (New York: Macmillan, 1986; reprint, Baltimore: Johns Hopkins University Press, 1997), 421–46; and Rennie B. Schoepflin, "Christian Science Healing in America," in *Other Healers: Unorthodox Medicine in America*, ed. Norman Gevitz (Baltimore: Johns Hopkins University Press, 1988), 192–214.

5. Gill, *Mary Baker Eddy*, 32–33. On the relationship between illness and creativity, see George Pickering, *Creative Malady: Illness in the Lives and Minds of Charles Darwin, Florence Nightingale, Mary Baker Eddy, Sigmund Freud, Marcel Proust, Elizabeth Barrett Browning* (New York: Oxford University Press, 1974), esp. chap. 11, "Mary Baker Eddy," 183–205.

6. Mary Baker Eddy, *Retrospection and Introspection* (Boston, 1916), 8–9.

7. Ibid., 13; Peel, *Mary Baker Eddy*, 1:23, 50–51.

8. See John D. Davies, *Phrenology—Fad and Science: A Nineteenth-Century American Crusade* (New Haven: Yale University Press, 1955); Madeleine B. Stern, *Heads and Headlines: The Phrenological Fowlers* (Norman: University of Oklahoma Press, 1971); Robert C. Fuller, *Mesmerism and the American Cure of Souls* (Philadelphia: University of Pennsylvania Press, 1982); Peter McCandless, "Mesmerism and Phrenology in Antebellum Charleston: 'Enough of the Marvellous,'"

Journal of Southern History 58 (1992): 199–230; R. Laurence Moore, *In Search of White Crows: Spiritualism, Parapsychology, and American Culture* (New York: Oxford University Press, 1977); and Mary Farrell Bednarowski, "Women in Occult America," in *The Occult in America: New Historical Perspectives*, ed. Howard Kerr and Charles L. Crow (Urbana: University of Illinois Press, 1983), 177–95.

9. Robert David Thomas, *"With Bleeding Footsteps": Mary Baker Eddy's Path to Religious Leadership* (New York: Alfred A. Knopf, 1994), 54–56.

10. Martin Kaufman, *Homeopathy in America: The Rise and Fall of a Medical Heresy* (Baltimore: Johns Hopkins University Press, 1971).

11. Mary Patterson to Phineas Parkhurst Quimby, 29 May 1862, as quoted in Thomas, *"With Bleeding Footsteps,"* 80.

12. Charles S. Braden, *Spirits in Rebellion: The Rise and Development of New Thought* (Dallas: Southern Methodist University Press, 1963), 89.

13. For a helpful discussion of the exchanges between Eddy and Quimby between 1863 and 1865, see Gill, *Mary Baker Eddy,* 137–46.

14. On the impact of Eddy's healing of Jarvis, see Peel, *Mary Baker Eddy,* 1:184–85, and Gill, *Mary Baker Eddy,* 629 n. 3.

15. Mary Baker Eddy, *Miscellaneous Writings, 1883–1896* (Boston, 1924), 24.

16. For a helpful view of the variety of similar ways nineteenth-century Americans blended medical and religious sectarianism, see Catherine L. Albanese, "Physic and Metaphysic in Nineteenth-Century America: Medical Sectarians and Religious Healing," *Church History* 55 (1986): 489–502, and Albanese, *Nature Religion in America: From the Algonkian Indians to the New Age* (Chicago: University of Chicago Press, 1990). For the ways writing provided some nineteenth-century women with an outlet, or even a weapon, see Ann D. Wood, "The 'Scribbling Women' and Fanny Fern: Why Women Wrote," *American Quarterly* 23 (1971): 3–24.

17. Letter from Mary M. Patterson to Julius Dresser, 15 February 1866; letter from J. A. Dresser to Mrs. Patterson, 2 March 1866, Archives of the Mother Church, Christian Science Center, Boston.

18. For a "spiritualized" view of Eddy's desire for economic independence, see Lyman P. Powell, *Mary Baker Eddy: A Life Size Portrait* (New York: Macmillan, 1930), 124–25.

19. Quotation from Peel, *Mary Baker Eddy,* 1:213.

20. The advertisement appeared on June 20 and 27 and July 4, 1868. See below the slightly different version that appeared in *Banner of Light* 26 (December 4, 1869): 5 and *Banner of Light* 26 (December 11, 1869): 7:

MARY M. B. GLOVER

Is prepared to take students at her residence, Stoughton, Mass., and teach them a SCIENCE by which all diseases are healed. Those who learn it are the greatest healers of the age. No medicine is used. All can learn it. No charges unless they can heal. Terms for payment settled one week after taking lessons.

21. "Questions and Answers in Moral Science," in *Mrs. Eddy as I Knew Her in 1870*, by Samuel P. Bancroft (Boston: Mary Beecher Longyear, 1875), 61–79.

22. Ibid., 75.

23. Quoted in Peel, *Mary Baker Eddy*, 1:239.

24. Ibid. Later the contract was reduced to 10 percent, and in 1876 it was "canceled by mutual agreement."

25. On the first year's income, see Powell, *Mary Baker Eddy*, 128. On the nature of the partnership and the size of Eddy's bank account, see Hugh A. Studdert Kennedy, *Mrs. Eddy: Her Life, Her Work and Her Place in History* (San Francisco: Farallon Press, 1947), 175; Powell, *Mary Baker Eddy*, 299 n. 65.

26. For the text of this letter, see Milmine, *Life of Mary G. Baker Eddy*, 58–59.

27. Bancroft, *Mrs. Eddy*, 85. The parallels between aspects of Christian Science and spiritualism have been noted by many scholars. For example, see Ann Braude, *Radical Spirits: Spiritualism and Women's Rights in Nineteenth-Century America* (Boston: Beacon Press, 1989), 182–89; R. Laurence Moore, "The Occult Connection? Mormonism, Christian Science, and Spiritualism," in *The Occult in America: New Historical Perspectives*, ed. Howard Kerr and Charles L. Crow (Urbana: University of Illinois Press, 1983), 135–61, and Ann Taves, *Fits, Trances, and Visions: Experiencing Religion and Explaining Experience from Wesley to James* (Princeton: Princeton University Press, 1999), 215–19.

28. "Spiritualism," as quoted in Bancroft, *Mrs. Eddy*, 88.

29. Milmine, *Life of Mary G. Baker Eddy*, 59.

30. See *The Science of Man, Embracing Questions and Answers in Moral Science* (1870, rev.), in Bancroft, *Mrs. Eddy*, 103–25; Mary Baker Glover, *The Science of Man, by Which the Sick Are Healed. Embracing Questions and Answers in Moral Science* (Lynn, Mass.: Thos. P. Nichols, 1870), 12–14. Two later editions of *The Science of Man* appeared, the first in 1879 shortly after Eddy's husband, Asa Gilbert Eddy, and her student Edward J. Arens had been tried for the murder of another former student, Daniel H. Spofford; the second in 1883, after the death of Asa and Eddy's claim that he had been "mesmerically poisoned" by the backslidden Arens. For Gill's speculation that there may have been a sexual dimension to the "rubbing," see Gill, *Mary Baker Eddy*, 196–208.

31. On the equal nature of the partnership, see Sibyl Wilbur, *The Life of Mary Baker Eddy* (New York: Concord, 1908), 194.

32. Mary Baker Eddy, *The First Church of Christ Scientist and Miscellany* (Boston: Allison V. Stewart, 1916), 114. After the sixth edition of 1883, the 1910 edition of *Science and Health* included *Key to the Scriptures*, a glossary containing allegorical meanings for key biblical words.

33. Although Eddy's interpretations bore her own indelible mark, the position of "metaphysical symbolism" and "poetic vocabulary" in American religious discourse had a long history. See Steven D. Cooley, "Applying the Vagueness of Language: Poetic Strategies and Campmeeting Piety in the Mid-Nineteenth Century," *Church History* 63 (1994): 582.

34. Glover, *Science of Man*, 5.

35. Mary Baker Eddy, *Science and Health with Key to the Scriptures* (Boston: First Church of Christ, Scientist, 1971), 465. Eddy believed that God exhibited both masculine and feminine characteristics, but with the exception of the third edition of *Science and Health* (1881), she regularly used masculine pronouns to refer to God.

36. Mary Baker Glover, *Science and Health* (Boston: Christian Science Publishing Company, 1875), 147.

37. Ibid., 341.

38. Ibid., 365.

39. See Gill, *Mary Baker Eddy*, 196–208. If Gill's speculation that the "rubbing" associated with the practices of some early Eddy followers had a sexual content, then there were further grounds for Eddy to take steps to ensure her movement's public reputation.

40. Mary B. Glover Eddy, *Science and Health*, 3d ed. (Lynn, Mass.: Asa G. Eddy, 1881), 1:203–4. The third through fifteenth editions (1881–85) each appeared in two volumes.

41. This contract and its signatories are quoted in Peel, *Mary Baker Eddy*, 1:286–87.

42. Norman Beasley, *Mary Baker Eddy* (New York: Duell, Sloan, and Pearce, 1963), 64–65; Gill, *Mary Baker Eddy*, 642 n. 29.

43. Glover, *Science and Health* (1875, 393; 1881, 1:89).

Chapter Two. Becoming a Practitioner and Teacher

Epigraph: Testimony of Mrs. Florence Williams, *Christian Science Journal* 9 (1891): 347.

1. See the testimony of "E. D. S.," *Christian Science Journal* 7 (January 1890): 516.

2. George B. Day, "Sheep, Shepherd, and Shepherdess," *Christian Science Journal* 5 (August 1887): 233.

3. On the percentage of female members, see Henry King Carroll, *The Religious Forces of the United States Enumerated, Classified, and Described* (New York: Charles Scribner's Sons, 1912), lvii. On the ratio of female to male practitioners, see Stephen Gottschalk, *The Emergence of Christian Science in American Religious Life* (Berkeley: University of California Press, 1973), 244, and Margery Q. Fox, "Power and Piety: Women in Christian Science" (Ph.D. diss., New York University, 1973), 143. Writing about the ratio of male to female practitioners in the 1880s and 1890s, Neal B. DeNood asserted that "fluctuating between one and two and one and five during the early years, the ratio . . . in the later years stabilized at approximately one to nine." See Neal B. DeNood, "The Diffusion of a System of Belief" (Ph.D. diss., Harvard University, 1937), as cited in Gage William Chapel, "Christian Science and the Nineteenth Century Woman's Movement," *Central States Speech Journal* 26 (1975): 148.

4. The female, urban, and middle-class nature of Christian Science membership has often been noted. For example, see Charles S. Braden, *Christian Science
Today: Power, Policy, Practice* (Dallas: Southern Methodist University Press,
1958), 7; Bryan Wilson, *Sects and Society: A Sociological Study of the Elim Tabernacle, Christian Science, and Christadelphians* (Berkeley: University of California
Press, 1961), 201–5; Robert Peel, *Mary Baker Eddy* (New York: Holt, Rinehart, and Winston, 1966–77), 2:162; and Gottschalk, *Emergence of Christian Science*, 257.

5. Penny Hansen, "Woman's Hour: Feminist Implications of Mary Baker
Eddy's Christian Science Movement, 1885–1910" (Ph.D. diss., University of California, Irvine, 1981), 16.

6. Eddy asserted on December 3, 1903: "You are not a Christian Scientist
until you do control the weather." See "Notes on the Divinity Course Given by
Mary Baker Eddy, the Discoverer and Founder of Christian Science, at Her
Home, Pleasant View, Concord, New Hampshire, during the Years Nineteen
Hundred Three, Nineteen Hundred Four, Nineteen Hundred Seven" (recorded
by Lida Fitzpatrick, C.S.D., 1933, Archives of the Mother Church, Christian
Science Center, Boston).

7. Mrs. M. A. Gaylord, "A Paper on Christian Science," *Christian Science
Journal* 12 (1894): 111–12.

8. Ezra W. Reid, "What Does Christian Science Reveal to Us To-day?"
Christian Science Journal 17 (1899): 481.

9. See Hansen, "Woman's Hour," 299–377.

10. On the connections between health reform, women, and self-help, see
Regina Markell Morantz, "Nineteenth Century Health Reform and Women: A
Program of Self-Help," in *Medicine without Doctors: Home Health Care in American
History*, ed. Guenter B. Risse, Ronald L. Numbers, and Judith Walzer Leavitt
(New York: Science History Publications/USA, 1977), 73–93.

11. [Mrs. Crosse or Mr. Wiggin?], "Slow and Rapid Healing," *Christian
Science Journal* 4 (1886): 150.

12. Eddy to Judge and Mrs. Hanna, 6 April 1898, vol. 39, no. 5218, 239,
Archives of the Mother Church, Christian Science Center, Boston.

13. Eddy to Archibald McLellan, 2 April 1905, vol. 25, no. 3097, 37, Archives
of the Mother Church, Christian Science Center, Boston.

14. Regarding the characteristics of the Progressive Era cultural ideal for
men, see E. Anthony Rotundo, *American Manhood: Transformations in Masculinity
from the Revolution to the Modern Era* (New York: Basic Books, 1993); Harry Brod,
ed., *The Making of Masculinities: The New Men's Studies* (Boston: Unwin Hyman,
1987); Joe L. Dubbert, *A Man's Place: Masculinity in Transition* (Englewood Cliffs,
N.J.: Prentice-Hall, 1979); Joe L. Dubbert, "Progressivism and the Masculinity
Crisis," *Psychoanalytic Review* 61 (1974): 443–55; and James R. McGovern, "David
Graham Phillips and the Virility Impulse of Progressives," *New England Quarterly* 39 (1966): 334–55.

15. Gottschalk, *Emergence of Christian Science*, 234.

16. For examples of those who have emphasized such religious motivations, see Robert Peel, *Christian Science: Its Encounter with American Culture* (New York: Henry Holt, 1958), and Gottschalk, *Emergence of Christian Science.*

17. Other than Hansen, those who have examined Christian Science testimonies include R. W. England, "Some Aspects of Christian Science as Reflected in Letters of Testimony," *American Journal of Sociology* 59 (1954): 448–53; Lucy Jayne Kamau, "Systems of Belief and Ritual in Christian Science" (Ph.D. diss., University of Chicago, 1971); Gottschalk, *Emergence of Christian Science,* 222–38; and Jean A. McDonald, "Mary Baker Eddy and the Nineteenth-Century 'Public' Woman: A Feminist Reappraisal," *Journal of Feminist Studies* 2 (1986): 105–11.

18. McDonald, "Mary Baker Eddy," 109. Not only did McDonald look at just one volume of the *Christian Science Journal,* but she chose the volume for 1910–11, the year of Eddy's death. If I am correct, by then Eddy's efforts to shift Christian Science from medical to religious roots was well under way, and the editors would have chosen testimonies to reflect that shift.

19. On "fads and 'isms,'" see J. M. Bradley, "Healing by Faith: Discussed from the Standpoint of a Physician," *Interstate Medical Journal* 13 (1906): 948; on "mental aberration," see Edward Huntington Williams, "The Christian Science Psychosis," *Medical Record* 96 (1919): 1048.

20. This was not unlike the influence of other female religious leaders such as Catherine Booth, cofounder of the Salvation Army, and Ellen G. White, cofounder of Seventh-day Adventism. See Normon H. Murdoch, "Female Ministry in the Thought and Work of Catherine Booth," *Church History* 53 (1984): 348–62, and Bertha Dasher, "Leadership Positions: A Declining Opportunity?" *Spectrum* 15 (December 1984): 35–37.

21. Barbara Welter, "The Feminization of American Religion: 1800–1860," in *Clio's Consciousness Raised: New Perspectives on the History of Women,* ed. Mary S. Hartman and Lois Banner (New York: Feminist Studies, 1974; reprint, New York: Octagon Books, 1976), 143.

22. Anne M. Boylan, "Evangelical Womanhood in the Nineteenth Century: The Role of Women in Sunday Schools," *Feminist Studies* 4 (1978): 62–80.

23. Carol Norton, "Woman's Cause: What the Work of the Founder of Christian Science Has Done for It," *Christian Science Journal* 13 (1895): 151–52.

24. For major contributions to our understanding of the reasons for the attraction of Americans to metaphysical healing, see Donald Meyer, *The Positive Thinkers: A Study of the American Quest for Health, Wealth and Personal Power from Mary Baker Eddy to Norman Vincent Peale* (New York: Random House, 1965; reissued under a slightly altered title in 1980); Gail Thain Parker, *Mind Cure in New England: From the Civil War to World War I* (Hanover, N.H.: University Press of New England, 1973); Robert C. Fuller, *Alternative Medicine and American Religious Life* (New York: Oxford University Press, 1989); and Mary Farrell Bednarowski, "Outside the Mainstream: Women's Religion and Women Religious Leaders in Nineteenth-Century America," *Journal of the American Academy of Religion* 48 (1980): 207–31.

25. Mary E. Collson, as quoted in Cynthia Grant Tucker, *A Woman's Ministry: Mary Collson's Search for Reform as a Unitarian Minister, a Hull House Social Worker, and a Christian Science Practitioner* (Philadelphia: Temple University Press, 1984), 67.

26. In the third edition of *Science and Health* (1881) Eddy even used feminine pronouns to refer to God. For other discussions of the themes touched on in this paragraph see Chapel, "Christian Science and the Nineteenth Century Woman's Movement," 142–49.

27. For helpful discussions of Shakers, see Stephen J. Stein, *The Shaker Experience in America: A History of the United Society of Believers* (New Haven: Yale University Press, 1992); Edward Deming Andrews, *The People Called Shakers: A Search for the Perfect Society*, new enl. ed. (New York: Dover, 1962); and Lawrence Foster, *Religion and Sexuality: Three American Communal Experiments of the Nineteenth Century* (New York: Oxford University Press, 1981), esp. chap. 2, "They Neither Marry nor Are Given in Marriage: The Origins and Early Development of Celibate Shaker Communities," 21–71.

28. On spiritualism in America, see Ann Braude, *Radical Spirits: Spiritualism and Women's Rights in Nineteenth-Century America* (Boston: Beacon Press, 1989), and R. Laurence Moore, *In Search of White Crows: Spiritualism, Parapsychology, and American Culture* (New York: Oxford University Press, 1977).

29. See the chapter titled "Marriage" in *Science and Health*.

30. Mary Baker Glover, *Science and Health* (Boston: Christian Scientist Publishing Company, 1875), 315, 316–17. Biographer Gillian Gill, *Mary Baker Eddy* (Reading, Mass.: Perseus Books, 1998), attributes much of the volatility of Eddy's relationship with the American public to her failure to adhere to the socially dominant norms regarding the participation of women in the public arena.

31. On the challenge of *The Woman's Bible* to the cultural assumptions underlying the notion of "woman's sphere," see Kathi L. Kern, "Rereading Eve: Elizabeth Cady Stanton and *The Woman's Bible*, 1885–1896," *Women's Studies* 19 (1991): 371–83.

32. Fox, "Power and Piety," 6–7.

33. Samuel A. Tannenbaum, "Eddyism, or Christian Science, Considered Medically, Legally and Economically," *American Medicine* 22 (1916): 44.

34. Tucker, *Woman's Ministry*, 68, 97–98. When Jean A. McDonald downplayed the preponderance of women in the movement, the importance of physical healings to their conversion, and the lure of becoming a practitioner and claimed that "the reason the movement attracted substantial numbers of both men and women undermines the interpretation of it as woman's quest for status and power in a male-dominated society," she failed to acknowledge that the knowledge Christian Scientists gained did not just answer religious dilemmas. Knowledge was also a means to achieve prestige, power, and money. See McDonald, "Mary Baker Eddy," 104.

35. Peel, *Mary Baker Eddy*, 2:10. For sources on the middle class, see Stuart M. Blumin, *The Emergence of the Middle Class: Social Experience in the American City,*

1760–1900 (New York: Cambridge University Press, 1989); Stuart M. Blumin, "The Hypothesis of Middle-Class Formation in Nineteenth-Century America: A Critique and Some Proposals," *American Historical Review* 90 (1985): 299–338; Mary P. Ryan, *Cradle of the Middle Class: The Family in Oneida County, New York, 1790–1865* (New York: Cambridge University Press, 1981); Peter Stearns, "The Middle Class: Toward a Precise Definition," *Comparative Studies in Society and History* 21 (1979): 377–96, 414–15; and Burton Bledstein, *The Culture of Professionalism: The Middle Class and the Development of Higher Education in America* (New York: Norton, 1976).

36. In one of the first attempts to give some texture to the amorphous American middle class, C. Wright Mills distinguished the "old" middle class—farmers, small retailers, artisans, and shopkeepers—from the "new" middle class—white-collar workers. See Mills, *White Collar: The American Middle Classes* (New York: Oxford University Press, 1951). On the occupational composition of Christian Scientists in 1910, see DeNood, "Diffusion of a System of Belief," as cited in Gottschalk, *Emergence of Christian Science*, 257.

37. Bledstein, *Culture of Professionalism*, 37, 113.

38. For an extended discussion of the ways work and community transformed the lives of single women, see Martha Vicinus, *Independent Women: Work and Community for Single Women* (Chicago: University of Chicago Press, 1985). For reflections on gender, class, and medical pluralism in America, see Hans A. Baer, "The American Dominative Medical System as a Reflection of Social Relations in the Larger Society," *Social Science and Medicine* 28 (1989): 1103–12.

39. For a firsthand account and analysis of female family members "doctoring" their households, see Emily K. Abel, "Family Caregiving in the Nineteenth Century: Emily Hawley Gillespie and Sarah Gillespie, 1858–1888," *Bulletin of the History of Medicine* 68 (1994): 573–99.

40. Barbara Ehrenreich and Deirdre English, *For Her Own Good: 150 Years of the Experts' Advice to Women* (Garden City, N.Y.: Anchor Press/Doubleday, 1978), 36. On the "cult of true womanhood," see Barbara Welter, "The Cult of True Womanhood: 1820–1860," *American Quarterly* 18 (1966): 151–74; Nancy F. Cott, *The Bonds of Womanhood: "Woman's Sphere" in New England, 1780–1835* (New Haven: Yale University Press, 1977); and Ann Douglas, *The Feminization of American Culture* (New York: Alfred A. Knopf, 1977).

41. For a defense grounded on the values inherent in the cult of true womanhood, see Barbara J. Harris, *Beyond Her Sphere: Women and the Professions in American History* (Westport, Conn.: Greenwood Press, 1978), 85.

42. Frances B. Cogan, *All-American Girl: The Ideal of Real Womanhood in Mid-Nineteenth-Century America* (Athens: University of Georgia Press, 1989), 9, 31, 18.

43. See Regina Markell Morantz, "The 'Connecting Link:' The Case for the Woman Doctor in Nineteenth-Century America," in *Sickness and Health in America*, 3d ed. rev., ed. Judith Walzer Leavitt and Ronald L. Numbers (Madison: University of Wisconsin Press, 1997), 213.

44. Gloria Moldow, *Women Doctors in Gilded-Age Washington: Race, Gender,*

and Professionalization (Urbana: University of Illinois Press, 1987), 6. Moldow also reports that "nationally, the number of trained women practitioners rose more than twelvefold between 1870 and 1900, from 544 in 1870 to 7,382 at the turn of the century."

45. For general discussions of the female attraction to and domestic nature of sectarian medicine, see Naomi Rogers, "Women and Sectarian Medicine," in *Women, Health, and Medicine in America: A Historical Handbook*, ed. Rima D. Apple (New York: Garland, 1990), 281–310; Morantz, "'Connecting Link,'" 117–28; Ronald L. Numbers, "Do-It-Yourself the Sectarian Way," in *Medicine without Doctors: Home Health Care in American History*, ed. Guenter B. Risse, Ronald L. Numbers, and Judith Walzer Leavitt (New York: Science History Publications/ USA, 1977), 49–72; John B. Blake, "Women and Medicine in Ante-Bellum America," *Bulletin of the History of Medicine* 39 (1965): 99–123.

46. Thomas J. Wolfe, "Every Woman a Physician: Women and Thomsonianism in Antebellum America," paper presented at the annual meeting of the American Association for the History of Medicine, Baltimore, Md., 1990.

47. Ronald L. Numbers, "The Making of an Eclectic Physician: Joseph M. McElhinney and the Eclectic Medical Institute of Cincinnati," *Bulletin of the History of Medicine* 47 (1973): 155–66; John S. Haller, *Medical Protestants: The Eclectics in American Medicine, 1825–1939* (Carbondale: Southern Illinois University Press, 1994); John S. Haller, *A Profile in Alternative Medicine: The Eclectic Medical College of Cincinnati, 1845–1942* (Kent, Ohio: Kent State University Press, 1999). On the history of Thomsonianism, see John S. Haller Jr., *The People's Doctors: Samuel Thomson and the American Botanical Movement, 1790–1860* (Carbondale: Southern Illinois University Press, 2000).

48. On the female connection to hydropathy, see Susan E. Cayleff, "Gender, Ideology, and the Water-Cure," in *Other Healers: Unorthodox Medicine in America*, ed. Norman Gevitz (Baltimore: Johns Hopkins University Press, 1988), 82–98; Susan E. Cayleff, *Wash and Be Healed: The Water-Cure Movement and Women's Health* (Philadelphia: Temple University Press, 1987); and Jane B. Donegan, *"Hydropathic Highway to Health": Women and Water-Cure in Antebellum America* (Westport, Conn.: Greenwood Press, 1986).

49. Anne Taylor Kirschmann, "Adding Women to the Ranks, 1860–1890: A New View with a Homeopathic Lens," *Bulletin of the History of Medicine* 73 (1999): 429–46. See also Numbers, "Do-It-Yourself," 58.

50. Penina Migdal Glazer and Miriam Slater, *Unequal Colleagues: The Entrance of Women into the Professions, 1890–1940* (New Brunswick, N.J.: Rutgers University Press, 1987), 83.

51. For example, see Moldow, *Women Doctors*, 6–7; Mary Roth Walsh, *"Doctors Wanted, No Women Need Apply": Sexual Barriers in the Medical Profession, 1835–1975* (New Haven: Yale University Press, 1977), 18, 42; and Blake, "Women and Medicine in Ante-Bellum America," 111–12.

52. William R. Rathvon, "The Practitioner in Business," *Christian Science Journal* 18 (1900): 290.

53. In the 1870s practitioners called themselves "Scientific Physicians"; after 1880 the term was usually "Christian Scientist."

54. After 1888 at the college, Eddy reduced this course from twelve lessons to seven, although other teachers continued to give twelve. Also, this degree was earlier simply Christian Scientist (C.S.) and indicated membership in the Christian Scientist Association, all members of which had been students of Eddy. See Peel, *Mary Baker Eddy*, 2:250, 371 n. 14; *Massachusetts Metaphysical College, Boston* (Cambridge, Mass.: John Wilson, 1884), 5; and *Journal of Christian Science* 2 (August 1884): 3.

55. [Mary B. Glover Eddy,] "Answers to Questions," *Journal of Christian Science* 1 (April 1883): 2.

56. J. H. W., "In the Class-Room," *Christian Science Journal* 4 (1886): 39.

57. Transcript of Proceedings, State of Nebraska v. E. M. Buswell, District Court for Gage County. Nebraska State Historical Society, Lincoln, Nebraska; RG 69, Nebraska Supreme Court, case 6495 (box 20); State v. Buswell, 100. 58 NW 728.

58. *Massachusetts Metaphysical College, Boston*, 3; document L15511, Archives of the Mother Church, Christian Science Center, Boston.

59. Peel, *Mary Baker Eddy*, 2:80–82, 111.

60. *Massachusetts Metaphysical College, Boston*, 3–4; *Curriculum of Massachusetts Metaphysical College, Boston*. Also, *Constitution and By-Laws of C.S.A.* (Published at the College, 1886), 6.

61. *Curriculum of Massachusetts Metaphysical College* (1886), 6–7.

62. See *Christian Science Journal* 3 (1886): 215. Ezra M. Buswell appears to have returned for the Normal course after only one year or so. Also, when he took the Normal course in 1888 he paid $200 in tuition. See Transcript of Proceedings, State of Nebraska v. E. M. Buswell, 100; 58 NW 728. Also, this degree was earlier called Christian Metaphysician (C.M.). See *Massachusetts Metaphysical College, Boston*, 5.

63. Apparently Eddy first taught a class specifically in metaphysical obstetrics in 1887. Nonetheless, earlier bulletins for her Massachusetts Metaphysical College indicated obstetrics as part of the curriculum. The 1882 bulletin identified Eddy as "Professor of Obstetrics, Metaphysics, and Christian Science"; the 1884 bulletin listed a full course in "Obstetrics" for $50; and the 1886 bulletin listed a course in "Mental and Physical Obstetrics" for $100. After 1901 the church authorized no special instruction in metaphysical obstetrics. See Peel, *Mary Baker Eddy*, 2:111; *Massachusetts Metaphysical College, Boston*, 6; *Curriculum of Massachusetts Metaphysical College* (1886), 6, 8.

64. Transcript of Proceedings, State of Nebraska v. E. M. Buswell, 101; 58 NW 728.

65. *Curriculum of Massachusetts Metaphysical College* (1886), 7. Also, this degree was earlier called Doctor of Christian Science (C.S.D.). See *Massachusetts Metaphysical College, Boston*, 5.

66. Gill, *Mary Baker Eddy*, 282, 646 n. 21.

67. Tyter attended Eddy's 10 January 1887 Primary class and her 30 October 1887 Normal class. See Ernest Sutherland Bates and John V. Dittemore, *Mary Baker Eddy: The Truth and the Tradition* (New York: Alfred A. Knopf, 1932), 467, 473.

68. The experiences of Josephine Tyter are detailed in her letter to Eddy, published in the *Christian Science Journal* 6 (1888): 145–46.

69. For a view of the experience of physicians in establishing a practice, see Charles E. Rosenberg, "Making It in Urban Medicine: A Career in the Age of Scientific Medicine," *Bulletin of the History of Medicine* 64 (1990): 163–86.

70. On the motives and experiences of women who migrated to turn of the century Chicago, see Joanne Meyerowitz, "Women and Migration: Autonomous Female Migrants to Chicago, 1880–1930," *Journal of Urban History* 13 (1987): 147–68.

71. On the use of "Christian Science Dispenser" as a title, see "Owen's Trial," *Davenport (Iowa) Democrat*, 6 June 1895, 1. On the insertion of business cards in local newspapers and the use of "Christian Science Teacher and Healer," see "May Live Forever," *Kalamazoo (Mich.) Daily Telegraph*, 12 February 1896, 1, 8.

72. "Christian Science a Success," *St. Joseph (Mo.) Daily Herald*, 12 February 1888, 5. Wheeler graduated from Eddy's 29 March 1886 Primary class and from her 7 February 1887 Normal class. See Bates and Dittemore, *Mary Baker Eddy*, 467, 472. Chicago practitioner S. A. Jefferson, C.S.B., placed several more sensational advertisements in the *Milwaukee Journal* during the spring of 1900. Among other things, he claimed that "over a million cures of disease in every form are now to the credit of Christian Science Healing" and that "you can be cured whether you believe in Christian Science or not." See "Christian Science Healing," *Milwaukee Journal*, 24 March 1900, 3; 12 May 1900, 9; 26 May 1900, 5.

73. Samual Putnam Bancroft, *Mrs. Eddy as I Knew Her in 1870* (Boston: Geo. Ellis, 1923), 22.

74. I. P. H., "Let the Voice of the National Association Be Heard," *Christian Science Journal* 8 (1890): 62.

75. Emma S. Davis, "Questions and Answers," *Christian Science Journal* 19 (1901): 475.

76. Ibid., 476.

77. Transcript of Proceedings, State of Nebraska v. E. M. Buswell, 89; 58 NW 728.

78. *Christian Science Journal* 5 (1887): 42, 203.

79. "Reminiscences of Mary Baker Eddy by Miss Julia S. Bartlett," quotation on 23; see also 14, 39–42, Archives of the Mother Church, Christian Science Center, Boston.

80. Julia Bartlett to Eddy, Wednesday night. The letter was answered by Eddy on 12 April 1884. Robert Peel dated Bartlett's letter 9 April 1884, Archives of the Mother Church, Christian Science Center, Boston.

81. Ary Johannes Lamme III, "The Spatial and Ecological Characteristics of the Diffusion of Christian Science in the United States: 1875–1910" (D.S.S. diss., Syracuse University, 1968); A. J. Lamme III, "Christian Science in the U.S.A., 1900–1910: A Distributional Study," Discussion Paper 3, Department of Geography, Syracuse University, April 1975.

82. Ferenc Morton Szasz, *The Protestant Clergy in the Great Plains and Mountain West, 1865–1915* (Albuquerque: University of New Mexico Press, 1988), 15–16. See also Eldon G. Ernst, "American Religious History from a Pacific Coast Perspective," in *Religion and Society in the American West: Historical Essays*, ed. Carl Guarneri and David Alvarez (New York: University Press of America, 1987), 3–39.

83. On Mormon missionary activities, see Samuel George Ellsworth, "A History of Mormon Missions in the United States and Canada, 1830–1860" (Ph.D. diss., University of California, 1951), esp. chap. 2, pp. 35–50; James B. Allen and Glen M. Leonard, *The Story of the Latter-day Saints* (Salt Lake City: Deseret Book Company, 1976), 48, 50, 73, 455–56; and Rex Thomas Price Jr., "The Mormon Missionary of the Nineteenth Century" (Ph.D. diss., University of Wisconsin–Madison, 1991), esp. chaps. 5 and 6, pp. 304–479. In 1893 the price for the cheapest edition of *Science and Health* rose from $3 to $3.18 per copy, although prices remained lower for bulk purchases.

84. Patricia R. Hill, *The World Their Household:The American Woman's Foreign Mission Movement and Cultural Transformation, 1870–1920* (Ann Arbor: University of Michigan Press, 1985), 25. On women and missions, see also Ann White, "Counting the Cost of Faith: America's Early Female Missionaries," *Church History* 57 (March 1988): 19–30; Ann Fagan, *This Is Our Song: Employed Women in the United Methodist Tradition* (Cincinnati: Women's Division of the General Board of Global Ministries, 1986); and Barbara Welter, "She Hath Done What She Could: Protestant Women's Missionary Careers in Nineteenth-Century America," *American Quarterly* 30 (1978): 624–38.

85. On the United States Census figures for 1890, see Carroll, *Religious Forces of the United States*, 382–83. I have depended on two main sources to arrive at the number 1,500. First, William Lyman Johnson, whose father William B. Johnson was one of the fourteen Boston practitioners who remained loyal to Eddy after the Corner schism in 1888, estimated that fewer than 1,000 of the 5,000 persons connected with Christian Science in Boston about 1887 were followers of Eddy. See his *History of the Christian Science Movement* (Brookline, Mass.: 1926), 2:253, as cited in Gottschalk, *Emergence of Christian Science*, 116. Second, William B. Johnson, the clerk of the church, reported at the 1906 annual meeting of the church that "the first annual meeting of the church was held in Chickering Hall, October 3, 1893, and the membership at that date was 1,545." See *Christian Science Sentinel* 8 (23 June 1906): 679.

86. The Christian Science membership figures of 82,332, gathered in the United States government's 1906 *Census of Religious Bodies* (Washington, D.C.:

U.S. Department of Commerce and Labor, Bureau of the Census, 1910), present their own difficulties. Efforts since the 1890 census, both within and outside the Eddy camp, to distinguish the "true" from the "pseudo" Christian Scientists (see chapter 4) made it likely that by 1906 only members of the Church of Christ, Scientist, would identify themselves as Christian Scientists to government census takers. However, since in 1906 most members held dual membership in the Mother Church and in their local churches, the census takers undoubtedly counted the membership of Boston's Mother Church twice in their calculations. I agree with Lamme that 3,000, not 41,634, is a fair estimate of the Mother Church's membership in 1906. With this adjustment, the census figures for Christian Science membership in 1906 should have been 43,690, which fits closely with the membership figures of 40,011 reported by the clerk, William B. Johnson, at the church's 1906 annual meeting. See Carroll, *Religious Forces of the United States*, lviii; Lamme, "Christian Science in the U.S.A.," 29 n. 25; *Christian Science Sentinel* 8 (23 June 1906): 679.

87. For Adventists and Mormons, see Carroll, *Religious Forces of the United States*, 465, 464.

88. Mary Baker Eddy, *Church Manual of the First Church of Christ, Scientist, in Boston, Massachusetts*, 73d ed. (Boston, 1925), 48; *Christian Science Sentinel* 15 (14 June 1913): 804; *Christian Science Journal* 43 (1925): 174; *Christian Science Journal* 52 (1934): 182.

89. The North Central region comprised the following states: North Dakota, South Dakota, Nebraska, Kansas, Minnesota, Iowa, Missouri, Wisconsin, Illinois, Indiana, Michigan, and Ohio.

90. Christian Scientist Josephine Curtis Woodbury believed she knew why residents of Denver, Colorado, displayed such a "universal interest" in Christian Science. According to her, sick people moved to Denver, hoping to live longer because of its healthful climate, and as their search for health continued, they discovered the healing power of Christian Science. See "The Queen City of the West," *Christian Science Journal* 6 (1888): 89–90.

91. Massachusetts (4,831), New York (5,671), Colorado (1,489), and California (2,753). In calculating these percentages I have used the adjusted 1906 census figures (see note 86).

92. Lamme, "Spatial and Ecological Characteristics," 46, 49, 55, 57, 58, 77–78, 85, 91. If, as these data suggest, there was a correlation between urbanism and the growth of Christian Science, then the case of Christian Science casts further doubt on the received wisdom that urbanization has always had a negative influence on American religiosity. For two recent contributions to this debate, see Roger Finke and Rodney Stark, "Religious Economies and Sacred Canopies: Religious Mobilization in American Cities, 1906," *American Sociological Review* 53 (1988): 41–49, and Kevin D. Breault, "New Evidence on Religious Pluralism, Urbanism, and Religious Participation," *American Sociological Review* 54 (1989): 1048–53.

Chapter Three. *"Occasions for Hope": Patients and Practitioners*

Epigraph: Report of Trial; State [and] Gardner T. Swarts v. David Anthony, case 18425, District Court of the Sixth Judicial District of the State of Rhode Island and Providence Plantations, 5–6. Phillips Memorial Library Archives, Providence College, Providence, Rhode Island.

1. "John H. Thompson, C.S.B.," *Quarterly News* (Longyear Historical Society and Museum) 22 (autumn 1985): 341.

2. Transcript of Proceedings, State of Nebraska v. E. M. Buswell, District Court for Gage County. Nebraska State Historical Society, Lincoln, Nebraska; RG 69, Nebraska Supreme Court, case 6495 (box 20), State v. Buswell, 102–3; 58 NW 728.

3. Ibid.

4. John H. Thompson Papers, 1908–15 (Manuscript Department, Duke University Library, Durham, North Carolina).

5. John H. Thompson to Mrs. Dove, 11 June 1910 (John H. Thompson Papers).

6. Mary Baker Eddy, *Science and Health with Key to the Scriptures*, (1910; reprinted, Boston: First Church of Christ, Scientist, 1934): 468.

7. John H. Thompson to Mr. Mulhall, 1 April 1910.

8. John H. Thompson to Mrs. Dove, 11 June 1910. For the source of this illustration by Thompson, see Eddy, *Science and Health* (1910), 128–29: "The addition of two sums in mathematics must always bring the same result. So is it with logic. If both the major and the minor propositions of a syllogism are correct, the conclusion, if properly drawn, cannot be false. So in Christian Science there are no discords nor contradictions, because its logic is as harmonious as the reasoning of an accurately stated syllogism or of a properly computed sum in arithmetic. Truth is ever truthful, and can tolerate no error in premise or conclusion."

9. John H. Thompson to Mrs. Dove, 11 June 1910. Eddy used the illustration of a dream numerous times. For a typical example, see Eddy, *Science and Health* (1910), 249–50.

10. John H. Thompson to Mr. McAuley, 21 December 1908.

11. John H. Thompson to Mrs. Dove, 22 July 1910. See also John H. Thompson to Mrs. Dove, 25 July 1910, which may be a second draft of the letter dated 22 July. The letter dated 25 July does not contain specific references to Dove's ailments or reference to the inclusion of a journal. It is not clear whether Dove received both letters or only the letter dated 25 July.

12. John H. Thompson to Mrs. Dove, 14 October 1910.

13. See, for example, Robert Peel, *Christian Science: Its Encounter with American Culture* (New York: Henry Holt, 1958); Stephen Gottschalk, *The Emergence of Christian Science in American Religious Life* (Berkeley: University of California Press, 1973); and Bryan Wilson, *Sects and Society: A Sociological Study of the Elim Tabernacle, Christian Science, and Christadelphians* (Berkeley: University of California Press, 1961), especially chaps. 6–10, pp. 121–215.

14. These contemporary studies include Walter I. Wardwell, "Christian Science Healing," *Journal for the Scientific Study of Religion* 4 (spring 1965): 175–81; Arthur Edmund Nudelman, "Christian Science and Secular Medicine" (Ph.D. diss., University of Wisconsin, 1970); Lucy Jayne Kamau, "Systems of Belief and Ritual in Christian Science" (Ph.D. diss., University of Chicago, 1971); Margery Q. Fox, "Power and Piety: Women in Christian Science" (Ph.D. diss., New York University, 1973). For three exceptions to this failure to examine historical practice, see Gottschalk, *Emergence of Christian Science*, esp. chap. 5, pp. 216–74; Margery Fox, "Conflict to Coexistence: Christian Science and Medicine," *Medical Anthropology* 8 (1984): 292–301; and Thomas C. Johnsen, "Christian Scientists and the Medical Profession: A Historical Perspective," *Medical Heritage* 2 (January–February 1986): 70–78.

15. On aging, see *Journal of Christian Science* 2 (August 1884): 3; on overweight, see *Journal of Christian Science* 2 (December 1884): 4.

16. Grant Wacker, "Marching to Zion: Religion in a Modern Utopian Community," *Church History* 54 (1985): 510.

17. Lloyd B. Coate, "Is Christian Science Christian?" *Christian Science Journal* 19 (1901): 27.

18. Transcript of Proceedings, People v. Willis V. Cole, 148 N.Y.S. 708. Transcript 1767, Library of the John Jay College of Criminal Justice, New York, 25–28.

19. Mary Baker Eddy, "Fallibility of Human Concepts," *Christian Science Journal* 7 (1889): 159–61.

20. Emma S. Davis, "Questions and Answers," *Christian Science Journal* 19 (1901): 480; emphasis supplied.

21. John H. Thompson to Mrs. King, 22 August 1910.

22. "Christian Scientists Held for Manslaughter," *New York Times*, 24 October 1902, 1.

23. Eddy made this remark on 28 May 1903. See "Notes on the Divinity Course Given by Mary Baker Eddy, the Discoverer and Founder of Christian Science, at Her Home, Pleasant View, Concord, New Hampshire, during the Years Nineteen Hundred Three, Nineteen Hundred Four, Nineteen Hundred Seven" (recorded by Lida Fitzpatrick, C.S.D., 1933), 13, Archives of the Mother Church, Christian Science Center, Boston.

24. Mary Baker G. Eddy, "Inconsistency," *Christian Science Journal* 24 (1906): 183.

25. See also A. C. Umbreit, ed., *Report of the Trial of Crecentia Arries and Emma Nichols, Charged with Practicing Medicine without a License, Before the Police Court of the City of Milwaukee, Giving a Verbatim Report of All the Testimony, Rulings of the Court, Arguments of Counsel and Decisions of the Presiding Judge* (Milwaukee: Edw. Keogh Press, 1900), 67.

26. John H. Thompson to Mr. McAuley, 15 December 1908.

27. "Mrs. Ward's Trial," *San Bernardino (Calif.) Daily Courier*, 5 February 1892, 3.

28. "Christian Scientists Held for Manslaughter," *New York Times*, 24 October 1902, 1.

29. Transcript of Proceedings, People v. Willis V. Cole, 88.

30. Mary Baker Glover, *Science and Health* (Boston: Christian Scientist Publishing Company, 1875), 290.

31. Arthur Corey, *Behind the Scenes with the Metaphysicians* (Los Angeles: DeVorss, 1968), 87–88.

32. Report of Trial; State [and] Gardner T. Swarts v. David Anthony, 5–6.

33. Jennie A. Spead v. Irving C. Tomlinson, New Hampshire Superior Court, Merrimack, ss, April Term, 1902. New Hampshire Supreme Court Library, Concord, New Hampshire; "Testimony of Mrs. Jennie A. Spead," *State of New Hampshire Supreme Court Briefs and Cases*, 226:302.

34. Report of Trial; State [and] Gardner T. Swarts v. David Anthony, 11–12.

35. See also Umbreit, *Report of the Trial of Crecentia Arries*, 47.

36. Edward Everett Norwood, "The Practice of Christian Science," *Christian Science Journal* 21 (1903): 151–52. For this view of treatment as a change in the practitioner, see also Willard S. Mattox, "What Is a Treatment?" *Christian Science Journal* 33 (1916): 689–91.

37. John H. Thompson to Mrs. Smith, 14 July 1910.

38. Transcript of Proceedings, State of Nebraska v. E. M. Buswell, 53–54; 58 NW 728.

39. H. P. S., " 'Faith Cure' Answered," *Christian Science Journal* 3 (1885): 56–57.

40. W. I. G[ill], "Conditions of Being Healed," *Christian Science Journal* 4 (1886): 205; Willis F. Gross, "In Due Season," *Christian Science Journal* 13 (1895): 15.

41. Willard S. Mattox, "To Beginners," *Christian Science Journal* 23 (1905): 501.

42. Mary Baker Eddy, "Questions and Answers," *Christian Science Journal* 3 (1885): 77.

43. "Answers to Questions," *Journal of Christian Science* 2 (December 1884): 4. This anonymous answer to a reader's question may have been written by the new editor of the *Journal*, Emma Hopkins, and not by Eddy. Either way, Eddy agreed that Christian Science healing never failed.

44. Transcript of Proceedings, State of Nebraska v. E. M. Buswell, 63; 58 NW 728.

45. Transcript of Proceedings, People v. Willis V. Cole, 80–81.

46. Jennie A. Spead v. Irving C. Tomlinson, 296.

47. John H. Thompson to Miss Renton, 6 April 1909.

48. John H. Thompson to Mr. McAuley, 21 December 1908.

49. Transcript of Proceedings, People v. Willis V. Cole, 30–31.

50. *Christian Science Instruction Recorded by Dr. Alfred E. Baker, M.D., C.S.D., during His Association with Mary Baker Eddy, Discoverer and Founder of Christian Science, Author of Its Textbook Science and Health with Key to Scriptures* (Boston:

J. Raymond Cornell, 1935), 12. The editors of this book give no dates for these instructions.

51. Transcript of Proceedings, State of Nebraska v. E. M. Buswell, 94–95; 58 NW 728. See also Umbreit, *Report of the Trial of Crecentia Arries*, 13.

52. Mary B. Glover Eddy, *Science and Health*, 3d ed. (Lynn, Mass.: Asa G. Eddy, 1881), 1:239.

53. Louise Delisle Radzinski, "Sympathy in Christian Science," *Christian Science Journal* 21 (1903): 274.

54. M. S. [Martha Sherman?], "Questions and Discussions," *Christian Science Journal* 7 (1889): 462.

55. Mary Baker Eddy, "Practitioners' Charges," *Christian Science Sentinel* 12 (1 January 1910): 350.

56. On this change in charges, see John H. Thompson to Mrs. King, 22 August 1910. In 1895 Mr. and Mrs. John R. Hatten, practitioners in Dayton, Ohio, regularly charged $5 a week, as did Mrs. H. E. Graybill of Atchison, Kansas. See "Investigation," *Dayton (Ohio) Daily Journal*, 23 March 1895, 1; *Atchison (Kans.) Daily Globe*, 16 May 1895, 2.

57. John H. Thompson to Mrs. Kelley, 26 August 1910.

58. John H. Thompson to Mr. A. F. Adams, 10 January 1911. This large bill doubtless reflected the seriousness of Adams's case. On 18 November 1911 Thompson had chastised Adams for not following his instructions and for not calling him immediately when he needed treatment. In Thompson's view, "When once met fully it [sickness] seldom returns for long at a time and finall[y] goes for good." See John H. Thompson to Mr. Adams, 18 November 1911.

59. James Neal Reminiscence; Julia Bartlett to Eddy, [9 April 1884]; Ursula N. Gestefeld to Eddy, 24 July 1884, all in Archives of the Mother Church, Christian Science Center, Boston. Transcript of Proceedings, People v. Willis V. Cole, 75. However, when the wife of practitioner Franklyn J. Morgan sued him for divorce, she claimed that he had an annual income from his practice of about $3,000. See "Irregularities of the Irregulars," *Illinois Medical Journal* 12 (1907): 388.

60. John H. Thompson to Mr. Alexander, 7 August 1913.

61. Transcript of Proceedings, State of Nebraska v. E. M. Buswell, 99, 93; 58 NW 728.

62. "Income of Washington Physicians," *National Medical Review* 3 (1895): 179. The questionnaire was sent to ten physicians, and eight responded: "The average professional income of the ten regular physicians of this city who have, in your opinion, the largest practices" was $9,500; "the average professional income of the next one hundred physicians" was $3,500; and "the average income of all the regular physicians of this city" was $2,000; Gloria Moldow, *Women Doctors in Gilded-Age Washington: Race, Gender, and Professionalization* (Urbana: University of Illinois Press, 1987), 121. For other evidence on income see ibid., 122; Mary Roth Walsh, *"Doctors Wanted, No Women Need Apply": Sexual Barriers in the Medical Profession, 1835–1975* (New Haven: Yale University Press, 1977), 183–84.

63. Ronald L. Numbers, "The Fall and Rise of the American Medical Profession," in *The Professions in American History*, ed. Nathan O. Hatch (Notre Dame, Ind.: University of Notre Dame Press, 1988), 62. For more detail, see Ronald L. Numbers, *Almost Persuaded: American Physicians and Compulsory Health Insurance, 1912–1920* (Baltimore: Johns Hopkins University Press, 1978).

64. James Neal Reminiscence; Eddy Correspondence (1883), both in Archives of the Mother Church, Christian Science Center, Boston. On the fees of orthodox physicians, see George Rosen, *Fees and Fee Bills: Some Economic Aspects of Medical Practice in Nineteenth Century America* (Baltimore: Johns Hopkins University Press, 1946).

65. "Editor's Table," *Christian Science Journal* 15 (1897): 517.

66. Wheeler v. Sawyer, 15 Atl. (Me.) 67 (1888); case file "George W. Wheeler versus Prince A. Sawyer," March Term, 1888, Supreme Judicial Court, Franklin County, Maine. Maine State Archives, Augusta, Maine.

67. On this distinction between payment for time or cure, see, for example, "The Marble Trial," *Sandusky (Ohio) Daily Register*, 20 August 1903, 3.

68. Transcript of Proceedings, State of Nebraska v. E. M. Buswell, 37–39; 58 NW 728.

69. Ibid., 78; 58 NW 728.

70. Ibid., 70; 58 NW 728.

71. Ibid., 92; 58 NW 728.

72. John H. Thompson to Mrs. Bulman, 1 July 1910.

73. "Compensation," *Christian Science Journal* 3 (1885): 92. See also E. M. T, "Recompense," *Christian Science Journal* 6 (1888): 267–68; Eva R. Wertz, "Just Recompense," *Christian Science Journal* 11 (1893): 313–17.

74. Kate Swope, "Diagnosis in Christian Science Practice, *Christian Science Journal* 21 (1903): 333.

75. Wacker, "Marching to Zion," 510.

76. Herbert S. Fuller, "Our Testimony Meetings," *Christian Science Journal* 16 (1898): 347.

77. "Thrill in Marble Case," *Sandusky (Ohio) Daily Register*, 15 August 1903, 3. For a recent example of culture clash over medicine, religion, and epilepsy, see Anne Fadiman, *The Spirit Catches You and You Fall Down: A Hmong Child, Her American Doctors, and the Collision of Two Cultures* (New York: Farrar, Straus, and Giroux, 1997).

78. C. S., "Instantaneous Healing," *Christian Science Journal* 10 (1892): 118–19.

79. *Christian Science Journal* 6 (1889): 563–65.

80. For example, see the testimonies of E. D. S. in *Christian Science Journal* 7 (1890): 516; N. A. E. in *Christian Science Journal* 7 (1890): 517–18; A. C. Eddy in *Christian Science Sentinel* (31 August 1899): 15; and Julia L. Nelson in *Christian Science Journal* 17 (1899): 440–41.

81. Transcript of Proceedings, State of Nebraska v. E. M. Buswell, 66–67; emphasis supplied; 58 NW 728.

82. Glover, *Science and Health* (1875), 400.

83. "The Mead Case," *Dayton (Ohio) Daily Journal*, 28 March 1895, 1.

84. Umbreit, *Report of the Trial of Crecentia Arries*, 90–91.

85. "The Ward Trial," *San Bernardino (Calif.) Daily Courier*, 4 February 1892, 3.

86. Ibid.

87. Mary Baker G. Eddy, *Science and Health with Key to the Scriptures* (Boston: Joseph Armstrong, 1905); "Answers to Questions," *Christian Science Journal* 2 (August 1884): 3. For a friendly discussion of this and subsequent concessions by Eddy to medicine as a result of her own ills, see Robert Peel, *Mary Baker Eddy* (New York: Holt, Rinehart, and Winston, 1966–77), 3:238–42.

88. John H. Thompson to Mr. Pike, 4 November 1909.

89. See Gillian Gill, *Mary Baker Eddy* (Reading, Mass.: Perseus Books, 1998), 400, 668 nn. 31, 32.

90. See also Umbreit, *Report of the Trial of Crecentia Arries*, 41.

91. "Healers Flock to Lathrop's Defence," *New York Evening Journal*, 24 October 1902, 5.

92. "Christian Scientists Held for Manslaughter," *New York Times*, 24 October 1902, 1.

93. "The Mead Case," *Dayton (Ohio) Daily Journal*, 28 March 1895, 1.

94. Quoted by Paul Schullery in "Hope for the Hook and Bullet Press," *New York Times Book Review*, 22 September 1985: 1, 34–35.

Chapter Four. Separating "True" Scientists from "Pseudo" Scientists

Epigraph: "Separation of the Tares and the Wheat," *Christian Science Journal* 6 (1889). 516.

1. Until July 1889 Christian Scientists often remained within their own denominations unless disfellowshiped.

2. Unless otherwise indicated, the account of the circumstances surrounding the death of Lottie A. James and her infant and the trial of Abby H. Corner is based on the following: "Fatal Faith," *Boston Daily Globe*, 21 April 1888, evening edition, 1; "Faith That Did Not Cure," *Boston Daily Globe*, 22 April 1888, morning edition, 2; "Law's Delay," *Boston Daily Globe*, 23 April 1888, evening edition, 1; "Medford's Victim," *Boston Daily Globe*, 24 April 1888, morning edition, 2; "Mrs. Corner in Court," *Boston Daily Globe*, 24 April 1888, evening edition, 5; and "Mrs. Corner Gets Bail," *Boston Daily Globe*, 25 April 1888, morning edition, 4.

3. For further information on Charles Cullis, see Raymond Joseph Cunningham, "Ministry of Healing: The Origins of the Psychotherapeutic Role of the American Churches" (Ph.D. diss., Johns Hopkins University, 1965), esp. chap. 1, pp. 1–44.

4. "Defence of Christian Science," *Boston Daily Globe*, 30 April 1888, morning edition, 3. This letter, signed "COMMITTEE ON PUBLICATION, Christian Science Association," was written by Mary Baker Eddy. The "committee" at the time consisted of Frank Mason, Josephine Curtis Woodbury, and Sarah Crosse.

5. Abraham J. Heschel, *The Prophets* (New York: Harper and Row, 1962), 5.

6. Bryan R. Wilson, *The Noble Savages: The Primitive Origins of Charisma and Its Contemporary Survival* (Berkeley: University of California Press, 1975), 5. See also Jonathan M. Butler, "Prophet or Plagiarist: A False Dichotomy," *Spectrum* 12 (1982): 44–48.

7. Minutes of the Christian Scientist Association, Archives of the Mother Church, Christian Science Center, Boston, as quoted in Robert Peel, *Mary Baker Eddy* (New York: Holt, Rinehart, and Winston, 1966–77), 2:95–96; Reminiscences of Mary Baker Eddy by Miss Julia S. Bartlett, Archives of the Mother Church, Christian Science Center, Boston; *Lynn (Mass.) Union,* 3 February 1882, as quoted in Peel, *Mary Baker Eddy,* 2:99.

8. *Massachusetts Metaphysical College, Boston* (Cambridge, Mass.: John Wilson, 1884), 16.

9. William Lyman Johnson, *The History of the Christian Science Movement* (Brookline, Mass., 1926), 2:253, cited in Stephen Gottschalk, *The Emergence of Christian Science in American Religious Life* (Berkeley: University of California Press, 1973), 116. On the names of the fourteen Boston practitioners who continued to advertise in the *Christian Science Journal,* see Norman Beasley, *The Cross and the Crown: The History of Christian Science* (London: George Allen and Unwin, 1953), 148.

10. L. M. M. [Luther M. Marston], *Mental Healing Monthly* 1 (1886): 53–54.

11. A. T. Buswell, "Human Leadership and Heavenly Liberty in Christian Science Culture," *Mental Healing Monthly* 2 (1887): 242–47.

12. M. B. G[age], *Mental Healing Monthly* 1 (1886): 8–9.

13. Also referred to as the Church of the Divine Unity (Scientist). See "Church Greeting," *Mental Healing Monthly* 3 (1888): 65.

14. L. M. M. [Luther M. Marston], *Mental Healing Monthly* 2 (1887): 252.

15. Peel, *Mary Baker Eddy,* 2:226.

16. Gail M. Harley, "Emma Curtis Hopkins: 'Forgotten Founder' of New Thought" (Ph.D. diss., Florida State University, 1991), 56; A. H. W. [Anne Holliday Webb], "Chicago and Its Early Workers," *Quarterly News* (Longyear Museum and Historical Society) 9 (autumn 1972): 138.

17. L. M. M. [Luther M. Marston], *Mental Healing Monthly* 1 (1886): 54, 37–38.

18. [Joseph Adams], *Chicago Christian Scientist* 1 (1887): 2. On Adams's ideological loyalty to Eddy, see his "Another Mile Stone," *Chicago Christian Scientist* 3 (1890): 266: "We do not buy, sell, or circulate any thing which claims to be Christian Science literature but what we have examined, and can endorse as being in harmony with the Bible and 'Science and Health.'"

19. Many biographical dictionaries give 1853 as Hopkins's date of birth. On the confusion regarding this date and the 1850 census report that confirms 1849, see Harley, "Emma Curtis Hopkins," 10–11.

20. *Christian Science* 2 (1890): 194.

21. *International Magazine of Christian Science* 3 (1888): 67. In October 1889 the *International Magazine of Christian Science* was renamed the *International Magazine of Truth* with a new editor, Alzire A. Chevaillier. See *International Magazine of Truth* 5 (1889): 26–30. In October 1890, angered by the selfishness of Christian Scientists, Chevaillier, the current editor, merged the *International Magazine of Truth* with Colville's the *Problem of Life*. See *International Magazine of Truth* 5 (1890): 461–64.

22. *Christian Science* was renamed *Universal Truth* sometime after 1888, but it continued publication until about 1900, when it merged with *Mind*. See Harley, "Emma Curtis Hopkins," 141.

23. Harley, "Emma Curtis Hopkins"; Charles S. Braden, *Spirits in Rebellion: The Rise and Development of New Thought* (Dallas: Southern Methodist University Press, 1963); Henry Warner Bowden, *Dictionary of American Religious Biography* (Westport, Conn.: Greenwood Press, 1977); and J. Gordon Melton, *Biographical Dictionary of American Cult and Sect Leaders* (New York: Garland, 1986), 116–17.

24. The *Mind Cure and Science of Life* became the *Mind Cure Journal* in December 1885; the latter in turn became the *Mental Science Magazine and Mind Cure Journal* in September 1887, and this in turn became the *Mental Science Magazine* in October 1887.

25. *Mind Cure and Science of Life* 1 (1885): 170–72; *Mind Cure and Science of Life* 1 (1885): 191; *Mind Cure and Science of Life* 1 (1885): 209–10; *Mind Cure Journal* 2 (1885): 44–45.

26. L. M. M. [Luther M. Marston], *Mental Healing Monthly* 1 (1886): 54–55.

27. A. J. Swarts, *Spiritual Healing Formula and Text Book* (Chicago: Author, 1887), 4–5.

28. Mary Baker G. Eddy, *Historical Sketch of Metaphysical Healing* (Boston: Author, 1885), 21. This pamphlet became the core of her later autobiography, *Retrospection and Introspection* (1891). For a highly suggestive if somewhat forced reading of *Retrospection*, see Stephen J. Stein, "*Retrospection and Introspection:* The Gospel according to Mary Baker Eddy," *Harvard Theological Review* 75 (1982): 97–116.

29. "Defence of Christian Science" first appeared in *Christian Science Journal* 2 (1885): 1–4; quotation on 4. This pamphlet provided the core for Eddy's later book *Christian Science: No and Yes* (1887).

30. This Christian Science effort to elevate the status of Eddy by highlighting her divine calling and prophetic gift paralleled the efforts of Sabbatarian Adventists earlier in the century to do the same for Ellen White. According to Godfrey T. Anderson, "Increasingly, leaders of the developing body of believers emphasized that acceptance of the prophetic counsel as a 'gift of the church' was important so that all in the fellowship of the Sabbatarian Adventist group would be in unity. By the time the church was ready to be officially organized in the early 1860s, J. N. Loughborough, D. T. Bourdeau, Uriah Smith, and other leading figures were suggesting that acceptance of the visions was essential to identify

fully with the group." See "Sectarianism and Organization, 1846–1864," in *Adventism in America*, ed. Gary Land (Grand Rapids, Mich.: William B. Eerdmans, 1986), 44. See also Ron Graybill, "The Power of Prophecy: Ellen White and the Women Religious Founders of the Nineteenth Century" (Ph.D. diss., Johns Hopkins University, 1983), esp. chap. 4, "Visions and Ecstasy," and Malcolm Bull and Keith Lockhart, *Seeking a Sanctuary: Seventh-day Adventism and the American Dream* (San Francisco: Harper and Row, 1989), 25–26.

31. H. P. S., "Perverted Metaphysics versus Theology of Christian Science," *Christian Science Journal* 3 (1886): 197.

32. "Questions and Answers," *Christian Science Journal* 3 (1886): 229. "Malicious mental malpractice" entailed using mental power to influence the thinking or behavior of others, for good or ill, without their consent. Such malpractice would involve the use of "malicious animal magnetism," also known as "MAM."

33. "Expositions, Wise and Otherwise," *Christian Science Journal* 4 (1886): 177, 178.

34. "Sunday Services on July Fourth," *Christian Science Journal* 4 (1886): 116; "Systematic Study," *Christian Science Journal* 4 (1886): 6.

35. See Judith Walzer Leavitt, *Brought to Bed: Childbearing in America, 1750–1950* (New York: Oxford University Press, 1986), quotation on 173. See also Richard W. Wertz and Dorothy C. Wertz, *Lying-In: A History of Childbirth in America* (New York: Schocken Books, 1977); Laurel Thatcher Ulrich, *A Midwife's Tale: The Life of Martha Ballard, Based on Her Diary, 1785–1812* (New York: Vintage Books, 1990), 162–203; Steven M. Stowe, "Obstetrics and the Work of Doctoring in the Mid-Nineteenth-Century American South," *Bulletin of the History of Medicine* 64 (1990): 540–66; and Charlotte G. Borst, *Catching Babies: The Professionalization of Childbirth, 1870–1920* (Cambridge: Harvard University Press, 1996).

36. *Christian Science Journal* 5 (1887): 395–97.

37. *Christian Science Journal* 5 (1887): 474. See also the testimony of Kittie Beck, *Christian Science Journal* 6 (1888): 148; the anonymous testimony in *Christian Science Journal* 6 (1889): 522; and the testimony of Dora Hossick in *Christian Science Journal* 6 (1889): 567.

38. *Christian Science Journal* 14 (1896): 349–50.

39. *Christian Science Journal* 15 (1897): 439.

40. The summary contained in this paragraph is based on a complete survey of the healing testimonies of the *Christian Science Journal* from 1883 to 1910. Physicians viewed things differently. Gustav F. Boehme Jr. recalled his attendance at one Christian Science delivery early in his career. The woman suffered a "median perineal tear well into the rectum, which I was not permitted to repair, 'For that tear is not a physical thing,' said the scientific interloper [practitioner]." See Gustav F. Boehme Jr., "Some Fallacies of Christian Science," *Medical Record* 99 (1921): 396.

41. Chester placed one of these colleges in San Francisco, where no "college" of Eddy's existed. This may identify Chester as a generic Christian Scientist.

The Emma Curtis Hopkins school had a strong contingent of followers in San Francisco.

42. The foregoing account is a reconstruction of events based on newspaper coverage of Agnes Chester's trial. See "Her Trial Begins," *Kalamazoo (Mich.) Daily Telegraph,* 11 February 1896, 1, and "May Live Forever," *Kalamazoo (Mich.) Daily Telegraph,* 12 February 1896, 1, 8.

43. Samuel L. Baker, "Physician Licensure Laws in the United States, 1865–1915," *Journal of the History of Medicine* 39 (1984): 173–97; "A Peculiar Case," *Christian Science Journal* 14 (1896): 34–35.

44. As quoted in Ernest Sutherland Bates and John V. Dittemore, *Mary Baker Eddy: The Truth and the Tradition* (New York: Alfred A. Knopf, 1932), 217.

45. Ibid.

46. Peel, *Mary Baker Eddy,* 2:111.

47. "Answers to Questions," *Christian Science Journal* 2 (December 1884): 4.

48. "M. H. P." [Mary H. Philbrick], *Christian Science Journal* 4 (1886): 123. Whether Philbrick changed the infant's position mentally or manually is unclear.

49. M. W. M. [Mary W. Munroe], *Christian Science Journal* 11 (1893): 278.

50. E. D. [*sic*] Greene, *Christian Science Journal* 5 (1887): 310.

51. G. B. Wickersham, *Christian Science Journal* 5 (1887): 311.

52. Letter of Eddy to Mrs. Annie V. C. Leavitt, 21 June 1887, vol. 88, no. 12966, 15, Archives of the Mother Church, Christian Science Center, Boston. Leavitt got into the December class, and according to Elizabeth Webster and Mary M. W. Adams, each of whom attended both of the 1887 classes, the class in December had been especially beneficial. See Elizabeth Webster, *Christian Science Journal* 6 (1888): 25; Mary M. W. Adams, *Christian Science Journal* 6 (1888): 76.

53. Mary Baker G. Eddy, "Take Notice," *Christian Science Journal* 4 (1887): 291. This notice was dated 17 February 1887. Before the obstetrics class authorized by the Board of Education of the Massachusetts Metaphysical College and taught by Alfred E. Baker, M.D., C.S.D., in June 1900, a similar notice appeared in the *Journal:* "Obstetrics is taught to those only who have received the College degrees"; Editor's Table, "The Massachusetts Metaphysical College," *Christian Science Journal* 18 (1900): 63.

54. "Defence of Christian Science," *Boston Daily Globe,* 30 April 1888, morning edition, 3.

55. Mary Baker Eddy, "Truth versus Error," *Christian Science Journal* 6 (1888): 319–20.

56. Mary Baker Eddy, "Christian Science Literature," *Christian Science Journal* 5 (1888): 633.

57. Peel, *Mary Baker Eddy,* 2:236–40.

58. C. A. Frye to Mr. Sawyer, June 7, [1888], document VO1111, Archives of the Mother Church, Christian Science Center, Boston.

59. Peel, *Mary Baker Eddy,* 2:250. Foster, who took the name Foster Eddy

after his adoption, graduated from the Hahnemann Medical College in Philadelphia in 1869. The editor of the *Christian Science Journal* also thought it worthy of note that Foster had received a certificate from Dr. W. W. Keen's Philadelphia School of Anatomy and had spent two years in clinical classes in two allopathic hospitals. See *Christian Science Journal* 11 (1893): 282.

60. In the tenth edition of the *Church Manual*, Eddy mandated that a teacher of obstetrics sit on the Board of Education. The board member had to have an M.D. degree, be "duly qualified to practise obstetrics," have a "diploma authorized by the State," and have at least a C.S.B. degree. See "Board of Education, article XXX, Organization and Rules. Sect. 3," in Mary Baker Eddy, *Church Manual*, 10th ed. (Boston: Christian Science Publishing Society, 1899), 65. The fourteenth edition (1899) required an examination: "A student in this class shall prepare a paper on *Accouchement*, giving in detail the physical and mental treatment requisite for the scientific and safe delivery of the mother. This paper shall be discussed by the class, and examined by the teacher, who shall decide as to the proper qualification of his pupils to practise obstetrics" (72). It also mandated no fewer than four lessons, one each day, a requirement that the seventeenth edition (1900) expanded to seven lessons. On the number and date of Baker's classes, see Peel, *Mary Baker Eddy*, 3:440 n. 45.

61. "Notes on Metaphysical Obstetrics: Used to Teach the Obstetric Class of June, 1900 of the Board of Education of the Massachusetts Metaphysical College. As Prepared by Dr. Alfred E. Baker, Teacher of Obstetrics" (reprint, n.d.). Baker's notes, as is often the case with lecture notes, appear to follow an outline that is not altogether consistent. Readers should not assume, therefore, that Baker discussed an issue in precisely the same place that my summary does.

62. Baker, "Notes on Metaphysical Obstetrics," 3–8.

63. Ibid., 18, 11–12.

64. Ibid., 8.

65. "Notes on the Divinity Course Given by Mary Baker Eddy, the Discoverer and Founder of Christian Science, at Her Home, Pleasant View, Concord, New Hampshire, during the Years Nineteen Hundred Three, Nineteen Hundred Four, Nineteen Hundred Seven" (recorded by Lida Fitzpatrick, C.S.D., 1933), 28, Archives of the Mother Church, Christian Science Center, Boston. Fitzpatrick dated this statement 7 December 1903. Eddy believed she "discovered" Christian Science in 1866, at which time she would have been nearing her forty-fifth birthday. She may have experienced menopause during the years surrounding 1866.

66. Baker, "Notes on Metaphysical Obstetrics," 9.

67. Ibid., 9–14, 32.

68. Ibid., 34–39.

69. Ibid., 39–41.

70. Ibid., 30.

71. Ibid., 29.

72. Ibid., 24.

73. Ibid., 21.

74. Ibid., 16–17 (abortion), 16 (father), 26 (mother-in-law), 21–22 (poverty), 22–23 (animal magnetism).

75. Ibid., 19–21, 24–25, 41, 9.

76. *Christian Science Hymnal* (Boston: Christian Science Publishing Society, 1937).

77. Baker, "Notes on Metaphysical Obstetrics," 33.

78. Ibid., 27–28. Locomotor ataxia, also called tabes dorsalis, is a progressive degeneration of the spinal cord and nerve roots, especially as a consequence of syphilis. Scrofula is primary tuberculosis of the lymphatic glands, especially of the neck.

79. "Amendments to Church By-Laws," *Christian Science Journal* 20 (1902): 38. See also Mary Baker Eddy, *Church Manual*, 25th ed. (Boston: Christian Science Publishing Society, 1901, 1902 printing), 20.

80. "The Present Hour," *Christian Science Journal* 7 (1889): 55.

81. She may have been aware that the Massachusetts legislature was on the verge of repealing the 1874 law under which the Massachusetts Metaphysical College and numerous other unorthodox educational institutions had been established. See Joel Whitney Tibbetts, "Women Who Were Called: A Study of the Contributions to American Christianity of Ann Lee, Jemima Wilkinson, Mary Baker Eddy and Aimee Semple McPherson" (Ph.D. diss., Vanderbilt University, 1976), 159 n. 62.

82. Mary B. G. Eddy, "The Way," *Christian Science Journal* 7 (1889): 431–34.

83. Mary Baker Eddy, "To the Christian World," *Christian Science Weekly* 1 (29 December 1898): 4–5.

84. For an insightful analysis of *Science and Health* as the locus of authority in Christian Science, see David L. Weddle, "The Christian Science Textbook: An Analysis of the Religious Authority of *Science and Health* by Mary Baker Eddy," *Harvard Theological Review* 84 (1991): 273–97. Similarly, Joseph Smith and early Mormons built authority on his writings; see Mario S. De Pillis, "The Quest for Religious Authority and the Rise of Mormonism," *Dialogue: A Journal of Mormon Thought* 1 (1966): 68–88.

85. Peel, *Mary Baker Eddy*, 3:90.

86. Braden, *Spirits in Rebellion*, 89–128.

Chapter Five. Physicians Debate Christian Science

Epigraphs: Mary J. Finley, "The Mind Cure," *Medical Record* 32 (1887): 590; Frank B. Wynn, "The Physician: Pathies, Isms and Cults in Medicine," *Journal of the Indiana State Medical Association* 14 (1921): 188.

1. For example, the author of "An Investigation of Christian Science," *Medical News* 74 (1899): 79, asserted that reportedly there were currently about thirty thousand Christian Scientists in the state of New York and one million in the United States. Samuel A. Tannenbaum stated that "it is said that there are now about one million [Christian Science] believers throughout the Christian world.

In the United States alone there are upward of four thousand healers." See his "Eddyism, or Christian Science: Considered Medically, Legally and Economically," *American Medicine* 22 (1916): 40.

2. For the statistics, see chapter 2, notes 85 and 86. Physicians attacked osteopaths and chiropractors during this same period; see Norman Gevitz, *The D.O.'s: Osteopathic Medicine in America* (Baltimore: Johns Hopkins University Press, 1982), and J. Stuart Moore, *Chiropractic in America: The History of a Medical Alternative* (Baltimore: Johns Hopkins University Press, 1993).

3. William A. White, "The Meaning of 'Faith Cures' and Other Extra-professional 'Cures' in the Search for Mental Health," *American Journal of Public Health* 4, n.s. (1914): 209.

4. Elisha Perkins (1741–99), a Connecticut physician, invented a "pair of small metal instruments, each about three inches long and both of the same shape, flat on one side, rounded on the other, and tapering from a hemispherical head to a sharp point. One of the tractors [as they were called] was gold in color, and the other silver." Perkins and others believed that when they were dragged across a patient's body in the correct way, pains and ailments disappeared. See James Harvey Young, *The Toadstool Millionaires: A Social History of Patent Medicines in America before Federal Regulation* (Princeton: Princeton University Press, 1961), 21–30, quotation on 24–25. On comparisons of Christian Science to Perkins's tractors, see Alexander B. Shaw, "The Medical Profession, Faith Cure, Christian Science and Mind Cure," *Transactions of the Medical Association of the State of Missouri*, 1889, 176–84; C. H. Hughes, "Christopathy and Christian Science (So-Called)," *Alienist and Neurologist* 20 (1899): 623; and "The End of Christian Science," *Western Medical Review* 5 (1900): 94–95. For a detailed effort to understand Christian Science as quackery, see Burnside Foster, "Rational Medicine and Charlatanry," *Transactions of the Minnesota State Medical Society*, 1894, 46–57.

5. Shaw, "Medical Profession," 176; James G. Kiernan, "Limitations of the Emmanuel Movement," *American Journal of Clinical Medicine* 16 (1909): 1089.

6. Henry K. Craig, "Christian Science Cures: Tales from Dreamland's Lore," *New York Medical Journal* 101 (1915): 514.

7. Edward Wallace Lee, "Psychic Healing, or, Properly Speaking, Treatment of Disease or Supposed Disease by Mental Influence," *New York Medical Journal and Philadelphia Medical Journal* 79 (1904): 97. For examples of authors who placed Christian Science within the history of faith healing, see J. M. Bradley, "Healing by Faith: Discussed from the Standpoint of a Physician," *Interstate Medical Journal* 13 (1906): 946–50, and Alfred D. Kohn, "Christian Science from a Physician's Standpoint," *Chicago Medical Recorder* 31 (1909): 376–97.

8. Finley, "Mind Cure," 590–93.

9. "An Aggressive Delusion," *JAMA* 33 (1899): 107.

10. For this view, see John B. Huber, "Faith Cures and the Law," *Medical Record* 59 (1901): 605–8, and R. Willman, "The Errors of Mind Healing," *Medical Herald* 29 (1910): 287–91.

11. For example, see "The 'Profitess' of Christian Science," *JAMA* 35 (1900): 829.

12. "Mrs. Mary Baker Eddy's Case of Hysteria," *JAMA* 48 (1907): 614–15; Ralph Wallace Reed, "A Study of the Case of Mary Baker G. Eddy," *Lancet-Clinic* 104 (1910): 360–65. Alfred D. Kohn called Eddy "an exceedingly neurotic woman, subject to hysterical seizures of all kinds" in his "Christian Science from a Physician's Standpoint," 378.

13. Edmund Andrews, "Christian Science: The New Theologico-philosophic Therapeutics," *JAMA* 32 (1899): 578.

14. John Harley Warner wrote that "the defining core of the proper physician's task became less the *exercise of judgment* and more the expert *application of knowledge.*" See *The Therapeutic Perspective: Medical Practice, Knowledge, and Identity in America, 1820–1885* (Cambridge: Harvard University Press, 1986), 260. On the antebellum medical ties between science and religion, see Steven M. Stowe, "Religion, Science, and the Culture of Medical Practice in the American South, 1800–1870," in *The Mythmaking Frame of Mind: Social Imagination and American Culture,* ed. James Gilbert et al. (Belmont, Calif.: Wadsworth, 1993): 1–24.

15. Bradley, "Healing by Faith," 946.

16. "Christian Science and Faith Healing," *Monthly Bulletin* (Iowa State Board of Health) 4 (1890–91): 162; James B. Taylor, *Christian Science Considered* (Cincinnati: Cranston and Stowe, 1888), 22. For other charges of blasphemy, see J. H. Richardson, "Christian Science," *Philadelphia Monthly Medical Journal* 1 (1899): 549–54, and Lee, "Psychic Healing," 97–102.

17. For example, see "Christian Science a Dangerous Fraud," *Illinois Medical Journal* 15 (1909): 546–49.

18. For example, see "Books on 'Christian Science,'" *JAMA* 51 (1908): 778

19. Lee, "Psychic Healing," 98; "Emmanuel Movement," *California State Journal of Medicine* 7 (1909): 79. The clergy had their own doubts about physicians, as the following comments reported by an osteopathic paper reveal: "Now and then I see where some Christian Scientists are arrested because a patient died under their treatment, and they are threatened with vengeance. In the name of common sense, what would become of the M.D.'s if they had to pay the penalty of graveyard subjects under their treatment? There would not be a doctor to-day out of the penitentiary or away from the gallows, and I am not bragging on Christian Scientists, nor have I any disposition to abuse them, for my father told me never to hit a cripple nor hurt a fool." See "Hot Roast of Doctors by the Rev. Sam P. Jones," *Cosmopolitan Osteopath* 3 (1900): 6.

20. On the absurdity of Christian Science, see H. V. Sweringen, "Christian Science," *Cincinnati Lancet-Clinic* 86, n.s., 47 (1901): 55–61; J. Playfair McMurrich, "A Philosophical, Anatomical, and Psychological Study of Christian Science: From an Anatomical Standpoint," *Medical Age* 21 (1903): 363–66.

21. "The Materiality of Germs and Automobiles," *JAMA* 72 (1919): 1469.

22. "Christian Science; The Greatest Delusion of the Age," *Journal of the Indiana State Medical Association* 14 (1921): 269. See also "A Test for 'Christian

Scientists,'" *JAMA* 34 (1900): 759, which reported that "a Detroit physician has been applying the *argumentum ad hominem* to the 'Christian Scientists,' by proposing to give them hypodermic injections of substances the effects of which on the system are known, and letting them try to nullify them by faith. This is a perfectly fair proposition, and if their faith really amounted to anything they ought to be willing to submit to such experimentation, but, so far, none have responded to the challenge."

23. *Minneapolis Journal*, 31 October 1904, 7.

24. Andrews, "Christian Science," 580.

25. "'Science,'" *JAMA* 69 (1917): 1267.

26. Kohn, "Christian Science from a Physician's Standpoint," 396.

27. "Eddyite Profits," *California State Journal of Medicine* 10 (1912): 452. See also "'Christian Science Healers Exempt from License Tax,'" *California State Journal of Medicine* 18 (1920): 288.

28. Andrews, "Christian Science," 580.

29. W. S. Turner, "Faith in Medicine," *Eclectic Medical Journal* 69 (1909): 598.

30. Tannenbaum, "Eddyism, or Christian Science," 41. The *Christian Science Journal* reproduced a report of the Agnes Chester case that appeared in the *St. Joseph (Mich.) Daily Herald*. In part the article stated that "Christian Science has become so popular in Kalamazoo as to interfere seriously with the incomes of the regular physicians of that city, and they have been aroused to the highest pitch of indignation at the course of affairs." See "A Peculiar Case," *Christian Science Journal* 14 (1896): 34–35.

31. John Ferguson, "False Systems of Healing: No. 1, Christian Science," *Canada Lancet* 50 (1916–17): 506.

32. J. E. Engstad, "The Doctrine of Christian Science a Theological, and Not a Medical, Problem," *Northwestern Lancet* 20 (1900): 364.

33. Tannenbaum, "Eddyism, or Christian Science," 39.

34. Andrews, "Christian Science," 578. See also Finley, "Mind Cure," 590.

35. "Psychic Healing and Christian Science," *New York Medical Journal and Philadelphia Medical Journal* 79 (1904): 377.

36. "Justice versus 'Science,'" *California State Journal of Medicine* 17 (1919): 424.

37. Andrews, "Christian Science," 581.

38. Taylor, *Christian Science Considered*, 44–45.

39. J. H. Richardson, "Christian Science," *Philadelphia Monthly Medical Journal* 1 (1899): 549–54; A. C. McClanahan, "A Shot at Christian Science from a Toy Pistol," *New York Medical Journal and Philadelphia Medical Journal* 81 (1905): 1222–25.

40. Kohn, "Physician's Standpoint," 387, 391.

41. "Another Faith Delusion," *JAMA* 37 (1901): 1245.

42. R. W. Kane, "By Their Fruits Ye Shall Know Them," *Christian Science Journal* 17 (1899): 339–40.

43. W. M. Polk, "The Medical Aspect of Christian Science," *New York Medical Journal* 73 (1901): 591–92.

44. Sweringen, "Christian Science," 57.

45. "Victims of 'Christian Science,' " *JAMA* 34 (1900): 633–34.

46. J. Playfair McMurrich, chair of anatomy at the University of Michigan, made just this point in "Philosophical, Anatomical, and Psychological Study of Christian Science," 365.

47. J. P. Widney, "Mind Cure," *Southern California Practitioner* 1 (1886): 176.

48. John B. Huber, "The Medical News' Investigation into the Claims of Christian Science," *Medical News* 74 (1899): 75–78, 107–10, 139–43, 168–70; quotations on 77, 170.

49. Henry H. Goddard, "The Effects of Mind on Body as Evidenced by Faith Cures," *American Journal of Psychology* 10 (1898–99): 431–502. The quotation appears on 444. For examples of favorable references to Goddard's study, see "The Genesis of a Miracle Cure," *Philadelphia Monthly Medical Journal* 41 (1899): 1048–49, and Robert T. Edes, "Mind Cures from the Standpoint of the General Practitioner," *Boston Medical and Surgical Journal* 151 (1904): 173–79.

50. Goddard, "Effects of Mind on Body," 435.

51. Ibid.

52. John M. Scudder, "Faith Cure—Christian Science," *Eclectic Medical Journal* 50 (1890): 506.

53. Richard C. Cabot, "One Hundred Christian Science Cures," *McClure's Magazine* 31 (1908): 472–76; Woodbridge Riley, Frederick W. Peabody, and Charles E. Humiston, *The Faith, the Falsity, and the Failure of Christian Science* (New York: Fleming H. Revell, 1925).

54. See "Christian Pseudoscience and Psychiatry," *Boston Medical and Surgical Journal* 143 (1900): 265; J. Leonard Corning, "The Limitations of the Influence of the Mind upon the Body," *Transactions of the Medical Society of the State of New York*, 1889, 347; Kohn, "Christian Science from a Physician's Standpoint," 387–88; Tom A. Williams, "Faith Cures and How They Act, Contrasted with the Principles of Scientific Mental Healing: with Case Reports," *Washington Medical Annals* 13 (1914): 317–22.

55. For examples of this view of the effects of religious and mind healing on organic disease, see J. P. Widney, "The Faith Cure Fallacy," *Southern California Practitioner* 1 (1886): 118–22; "Christian Science Illustrated," *Monthly Bulletin* (Iowa State Board of Health) 6 (1892–93): 8; T. F. Lockwood, "Faith, Fraud and Suggestive Therapeutics," *Transactions of the Medical Association of the State of Missouri*, 1899, 197–208; J. M. Bradley, "Healing by Faith: Discussed from the Standpoint of a Physician," *Interstate Medical Journal* 13 (1906): 946–50; and Tannenbaum, "Eddyism, or Christian Science," 39–48.

56. For helpful views of the psychological developments of the late nineteenth and early twentieth centuries and their relation to religion, see John Chynoweth Burnham, *Psychoanalysis and American Medicine, 1894–1918: Medicine,*

Science, and Culture, Psychological Issues, Monograph 20 (New York: International Universities Press, 1967); Nathan G. Hale Jr., *Freud and the Americans: The Beginnings of Psychoanalysis in the United States, 1876–1917* (New York: Oxford University Press, 1971); Eric Caplan, *Mind Games: American Culture and the Birth of Psychotherapy* (Berkeley: University of California Press, 1998); and Eugene Taylor, *Shadow Culture: Psychology and Spirituality in America* (Washington, D.C.: Counterpoint, 1999).

57. For example, see Corning, "Limitations of the Influence of the Mind upon the Body," 340–49; A. T. Conley, "Placeboes," *Transactions of the Minnesota State Medical Society*, 1891, 12–17; W. P. Hartford, "Subjective Therapeutics," *Medical Record* 54 (1898): 157–59; Sweringen, "Christian Science," 55–61; Edes, "Mind Cures from the Standpoint of the General Practitioner," 173–79; and Lee, "Psychic Healing," 97–102.

58. Freudian convert Samuel A. Tannenbaum offered Freud's theory of transference as an explanation in "Eddyism, or Christian Science," 39–48. Freudian critic Tom A. Williams suggested Pavlovian conditioning as a psychological explanation for mental healing in "Faith Cures and How They Act," 317–22. For histories of the psychotherapy movement, see Caplan, *Mind Games*, and Burnham, *Psychoanalysis and American Medicine*, 73–81.

59. Goddard, "Effects of Mind on Body," 445, 481, 501.

60. Regarding the former see, for example, H. Gasser, "The Science of Christian Science," *Transactions of the State Medical Society of Wisconsin* 33 (1899): 338–47, and Edes, "Mind Cures from the Standpoint of the General Practitioner," 173–79. Regarding the latter, see W. P. Hartford, "Subjective Therapeutics," *Medical Record* 54 (1898): 157–59.

61. Robert M. Wenley, "A Philosophical, Anatomical, and Psychological Study of Christian Science: From a Philosophical Standpoint," *Medical Age* 21 (1903): 361. For similar views, see also "Christian Science and Primitive Christianity," *New York State Journal of Medicine* 8 (1908): 500–501.

62. Engstad, "Doctrine of Christian Science," 364, 363.

63. For an example of these sentiments, see "Christian Science and Primitive Christianity," 500–501.

64. Hildegarde H. Longsdorf, "Christian Science and Its Relation to the Medical Profession," *Transactions of the Medical Society of the State of Pennsylvania* 25 (1894): 156.

65. D. A. Richardson, "Christian Science versus Medical Science," *Denver Medical Times* 16 (1896–97): 351.

66. Ibid., 353.

67. Ibid., 356.

68. Warner, *Therapeutic Perspective*, 238–39.

69. "Christian Science and Nihilistic Therapeutics," *Medical News* 73 (1898): 556–57. These sentiments were echoed by Ohio eclectic physician W. S. Turner in "Faith in Medicine," *Eclectic Medical Journal* 69 (1909): 599, where he wrote that we should "make our study along the line of specific medication, by which

disease-expressions as manifested by symptoms form the basis of our therapeutic application."

70. "Faith Cure, Christian Science, etc.," *JAMA* 35 (1900): 304.

71. "The Death of Mrs. Eddy," *JAMA* 55 (1910): 2068.

72. Shaw, "Medical Profession," 182.

73. For similar views, see C. A. F. Lindorme, "The Christian-Science Bubble," *Atlanta Journal–Record of Medicine* 2 (1900–1901): 318, and J. M. Aikin, "Mental Therapeutics in Medicine," *Medical News* 83 (1903): 494–95.

74. P. Maxwell Foshay, "Why Do the Quack and the Faith Curist at Times Succeed," *JAMA* 35 (1900): 637–38.

75. D. W. Harrington, "Mind and Medicine," *American Medicine* 15 (1909): 558.

76. Ibid., 561.

77. Aikin, "Mental Therapeutics in Medicine," 494.

78. Taylor, *Christian Science Considered*, 33.

79. A. W. Abbott, "President's Address," *Transactions of the Minnesota State Medical Society*, 1893, 6.

80. Longsdorf, "Christian Science and Its Relation to the Medical Profession," 151.

81. R. L. T. [R. L. Thomas], "Faith Cure," *Eclectic Medical Journal* 59 (1899): 622.

82. As St. Joseph, Missouri, physician A. F. Stephens put it, "Confidence in the doctor counts a great deal in our treatment of the sick[.] If we can make the patient feel confident that we can relieve him, the work is half accomplished. If not we have a hard, slow task." See his "Christian Science," *Eclectic Medical Journal* 53 (1893): 601.

83. L. C. Mitchell, "The Influence of Mind in Disease," *Northwestern Lancet* 5 (1885): 27–28; E. W. Mitchell, "Faith-Cure, Mind-Cure, and Christian Science," *Cincinnati Lancet-Clinic* 59, n.s., 20 (1888): 411; Shaw, "Medical Profession," 184.

84. For further discussions of the Emmanuel Movement, see Robert Bruce Mullin, "The Debate over Religion and Healing in the Episcopal Church: 1870–1930," *Anglican and Episcopal History* 60 (1991): 213–34; Raymond J. Cunningham, "From Holiness to Healing: The Faith Cure in America," *Church History* 43 (1974): 499–513; Raymond Joseph Cunningham, "The Emmanuel Movement: A Variety of American Religious Experience," *American Quarterly* 14 (1962): 48–63; Raymond Joseph Cunningham, "Ministry of Healing: The Origins of the Psychotherapeutic Role of the American Churches" (Ph.D. diss., Johns Hopkins University, 1965), esp. chaps. 4–5, pp. 113–89; and Sanford Gifford, "Medical Psychotherapy and the Emmanuel Movement in Boston, 1904–1912," in *Psychoanalysis, Psychotherapy and the New England Medical Scene, 1894–1944*, ed. George E. Gifford Jr. (New York: Science History Publications/USA, 1978), 106–18.

85. Richard C. Cabot, "Mind Cure: Its Service to the Community," *Colorado Medicine* 4 (1907): 5.

86. Ibid., 7.

87. On the withdrawal of Cabot and Putnam, see Hale, *Freud and the Americans*, 248. For views similar to Putnam's see [L.] Watkins, "Faith Healing," *Eclectic Medical Journal* 69 (1909): 260–61. D. W. Harrington, president of the Brainerd Medical Society, Milwaukee, Wisconsin, even wondered about the church's motives in becoming interested in psychotherapeutics. He wrote that "neither is it to the credit of the Christian denominations that they should be so anxious to establish church clinics now that Christian Science has become such a financial success." See his "Mind and Medicine," *American Medicine* 15 (1909): 558.

88. "Psychotherapy from a Medical and Religious Standpoint," *Journal of the Minnesota State Medical Association and the Northwestern Lancet* 28 (1908): 233. For further support of this view, see "Discussion of the Symposium on Psychotherapy," *Illinois Medical Journal* 15 (1909): 512–17; Reinhold Willman, *The Errors of Mind Healing Compared with the Miracles of Christ and His Disciples in the Healing of the Afflicted as Viewed by a Physician* (St. Joseph, Mo.: Advocate, 1909), 162.

89. See Hale, *Freud and the Americans*, 248.

90. Nathan G. Hale Jr., ed., *James Jackson Putnam and Psychoanalysis* (Cambridge: Harvard University Press, 1971), 14; William G. Rothstein, *American Medical Schools and the Practice of Medicine: A History* (New York: Oxford University Press, 1987), 155–56; and Jeanne L. Brand, "Neurology and Psychiatry," in *The Education of American Physicians: Historical Essays*, ed. Ronald L. Numbers (Berkeley: University of California Press, 1980), 229–40.

91. William A. White, "The Meaning of 'Faith Cures' and Other Extraprofessional 'Cures' in the Search for Mental Health," *American Journal of Public Health* 4, n.s. (1914): 216.

92. Burnham, *Psychoanalysis and American Medicine*, 78–79; Hale, *James Jackson Putnam*, 16.

93. Some evidence suggests that the secularization of medicine may have led Americans to become more willing to bring malpractice suits against physicians rather than leaving misfortune to providence. See Kenneth Allen De Ville, *Medical Malpractice in Nineteenth-Century America: Origins and Legacy* (New York: New York University Press, 1990), 114–37.

Chapter Six. Therapeutic Choice or Religious Liberty?

Epigraphs: Samuel A. Tannenbaum, "Eddyism, or Christian Science: Considered Medically, Legally and Economically," *American Medicine* 22 (1916): 48; M. [Willard S. Mattox], "Editor's Table," *Christian Science Journal* 20 (1903): 773–74.

1. Unless otherwise indicated, this account of the circumstances surrounding the trial of Harriet Evans is based on the following: "Ridiculed," *Cincinnati Enquirer*, 31 October 1898, 5; "Christian Science," *Cincinnati Enquirer*, 7 November 1898, 5; "Precedent May Now Be Established . . . ," *Cincinnati Enquirer*, 12 November 1898, 8; "Succumbed to Typhoid Did McDowell, over Whom the Regu-

lars Contended against Christian Scientists," *Cincinnati Enquirer*, 14 November 1898, 6; "Pressed by the State Board Will Be Case against Christian Scientist Evans," *Cincinnati Enquirer*, 15 November 1898, 5; "State Board," *Cincinnati Enquirer*, 15 November 1898, 5; "Prayers Did Not Save McDowell," *Cincinnati Enquirer*, 16 November 1898, 12; "Neglect Caused Death of Thomas McDowell . . . ," *Cincinnati Enquirer*, 19 November 1898, 16; "Christian Science Attacked by Rev. Charles Sutton, of Newport," *Cincinnati Enquirer*, 21 November 1898, 6; "Guilty," *Cincinnati Enquirer*, 9 December 1898, 10; "Reversed," *Cincinnati Enquirer*, 31 December 1898, 2; "More Christian Science," *New York Times*, 12 November 1898, 6; "Christian Science Inquiry," *New York Times*, 16 November 1898, 1; [Editorial], "Topics of the Times," *New York Times*, 24 December 1898, 6; 9 Ohio Decisions 222 (1889 [*sic*]).

2. "More Christian Science," *New York Times*, 12 November 1898, 6.

3. The words "he became rapidly worse" were those used by Judge J. Hollister in his decision overturning Evans's initial conviction. See 9 Ohio Decisions 222 (1889 [*sic*]). When the editors of the *Christian Science Journal* summarized the case they reported that "His physicians gave no relief. . . ." See *Christian Science Journal* 16 (1899): 827.

4. *Christian Science Journal* 16 (1899): 827.

5. In clear contrast to Judge J. Hollister's view that McDowell's condition declined, the editors of the *Christian Science Journal* reported that Evans's treatment "was done with noticeable improvement for five days." See *Christian Science Journal* 16 (1899): 827.

6. "Guilty, Was the Jury's Verdict," *Cincinnati Enquirer*, 22 December 1898, 12.

7. 9 Ohio Decisions 222 (1889 [*sic*]).

8. *Christian Science Journal* 16 (1899): 827–35.

9. "Editor's Table," *Christian Science Journal* 13 (1895): 84–85.

10. Raymond J. Cunningham, "The Impact of Christian Science on the American Churches, 1880–1910," *American Historical Review* 72 (1967): 885–905.

11. For example, see Robert Bruce Mullin, *Miracles and the Modern Religious Imagination* (New Haven: Yale University Press, 1996); Robert C. Fuller, *Alternative Medicine and American Religious Life* (New York: Oxford University Press, 1989); and Donald Meyer, *The Positive Thinkers: A Study of the American Quest for Health, Wealth and Personal Power from Mary Baker Eddy to Norman Vincent Peale* (Garden City, N.Y.: Doubleday, 1965; reissued under a slightly altered title in 1980).

12. Osteopaths and chiropractors also struggled with the efforts of medical doctors to control the practice of medicine. For the ways they negotiated medical licensing, see Norman Gevitz, *The D.O.'s: Osteopathic Medicine in America* (Baltimore: Johns Hopkins University Press, 1982), and J. Stuart Moore, *Chiropractic in America: The History of a Medical Alternative* (Baltimore: Johns Hopkins University Press, 1993).

13. Many of these cases were reported in the *Christian Science Journal*.

14. For an overview of the context for these efforts by physicians, see Ronald L. Numbers, "The Fall and Rise of the American Medical Profession," in *The Professions in American History*, ed. Nathan O. Hatch (South Bend, Ind.: University of Notre Dame Press, 1988), 51–71.

15. For an overview of medical licensing in America during this period, see Samuel L. Baker, "Physician Licensure Laws in the United States, 1865–1915," *Journal of the History of Medicine* 39 (1984): 173–97, and William G. Rothstein, *American Physicians in the Nineteenth Century: From Sects to Science* (Baltimore: Johns Hopkins University Press, 1967), 305–10.

16. For the influence of educational reforms, see Richard Harrison Shryock, *Medical Licensing in America, 1650–1965* (Baltimore: Johns Hopkins University Press, 1967), 47–60. For the influence of scientific medicine and the merger of the sects, see Rothstein, *American Physicians*, 298–326, and E. Richard Brown, *Rockefeller Medicine Men: Medicine and Capitalism in America* (Berkeley: University of California Press, 1979), 88–91.

17. For the view that physicians sought a competitive advantage through licensing, see Paul Starr, *The Social Transformation of American Medicine* (New York: Basic Books, 1982), 102–12. For the view that they sought a monopoly, see Rothstein, *American Physicians*, 305–10. To view medical licensure in the general context of occupational licensing, see Lawrence M. Friedman, "Freedom of Contract and Occupational Licensing, 1890–1910: A Legal and Social Study," *California Law Review* 53 (1965): 487–534. For a helpful view of the role that both occupational power and state governments played in the expansion of occupational licensing, see Xueguang Zhou, "Occupational Power, State Capacities, and the Diffusion of Licensing in the American States: 1890 to 1950," *American Sociological Review* 58 (1993): 536–52. See also Samuel Haber, *The Quest for Authority and Honor in the American Professions, 1750–1900* (Chicago: University of Chicago Press, 1991), 319–58.

18. Samuel L. Baker, "A Strange Case: The Physician Licensure Campaign in Massachusetts in 1880," *Journal of the History of Medicine* 40 (1985): 286–308.

19. Dent v. State of West Virginia, 9 Sup. Ct. Rep. 231 (1888). For a view of the "judicial review of economic and social legislation," including the Dent case, between 1890 and 1910, see Friedman, "Freedom of Contract," 487–534.

20. Mary B. G. Eddy, "Prayer and Healing," *Journal of Christian Science* 2 (February 1885): 5; Emma Hopkins, "Fiat Justitia," *Journal of Christian Science* 2 (November 1884): 4.

21. [Mary B. Glover Eddy,] "Answers to Questions," *Journal of Christian Science* 1 (April 1883): 2.

22. "Legal Persecutions," *Christian Science Journal* 4 (1886): 221.

23. See "Little Rhody," *Christian Science Journal* 5 (1887): 156, and "Medical Monopoly Not Wanted," *Christian Science Journal* 7 (1889): 21–22.

24. J[ohn] F. Linscott, "Scientific Organization," *Christian Science Journal* 7 (1889): 398.

25. C. H. Hughes, "Christopathy and Christian Science (So-Called)," *Alienist and Neurologist* 20 (1899): 621.

26. [Editorial], "Punish the Impostors," *New York Times*, 25 November 1898, 4.

27. Charlotte Eddy Post (apparently unrelated to Mary Baker Eddy) may have been the first Christian Scientist brought to trial for her practices. According to Mary Baker Eddy, "My book Science and Health was sent to the Judge and when the case came before Court the Judge dismissed it by saying ['I find you have no case, but God has one[.'] Wasn't that fine?" See Mary Baker Eddy to Mr. Wiggin, 16 April 1887, vol. 18, no. 2189, 35, Archives of the Mother Church, Christian Science Center, Boston. Regarding the intended influence of the Agnes Chester trial, see "A Peculiar Case," *Christian Science Journal* 14 (1896): 34–35.

28. Buswell also pastored in the nearby communities of Blue Springs (twenty-eight members) and Weeping Water (membership unknown). On the Buswells, see A. H. W. [Anne H. Webb], "Pioneers in the West," *Quarterly News* (Longyear Historical Society) 9 (spring 1972): 130–32; "Elizabeth Buswell, C.S.D." and "Ezra M. Buswell, C.S.D.," in *Pioneers in Christian Science* (Brookline, Mass.: Longyear Historical Society, 1972, rev. 1980); and William Holman Jennings, "Christian Science," in *Illustrated History of Nebraska*, ed. J. Sterling Morton, Albert Watkins, and George L. Miller (Lincoln: Jacob North, 1907), 2:475–80.

29. Eddy to Buswell, 27 February 1893, quoted in *Christian Science Journal* 11 (1893): 85–86. For information on the Stewarts' trial, see the sources cited in the bibliography.

30. "The Buswell Trial," *Beatrice (Nebr.) Daily Express*, 2 March 1893, 1.

31. "He Is Not Guilty," *Beatrice (Nebr.) Daily Express*, 3 March 1893, 1.

32. Transcript of Proceedings, State of Nebraska v. E. M. Buswell, District Court for Gage County. Nebraska State Historical Society, Lincoln, Nebraska; RG 69, Nebraska Supreme Court, case 6495 (box 20), State v. Buswell, 99, 117; "The Buswell Trial," *Beatrice (Nebr.) Daily Express*, 2 March 1893, 1.

33. "The Buswell Trial," *Beatrice (Nebr.) Daily Express*, 2 March 1893, 1.

34. See "Rev. Buswell on Trial," *Beatrice (Nebr.) Daily Times*, 28 February 1893, 4; "They Cured by Faith," *Beatrice (Nebr.) Daily Times*, 1 March 1893, 4; "In the Jury's Hands," *Beatrice (Nebr.) Daily Times*, 2 March 1893, 4; "Buswell Acquitted," *Beatrice (Nebr.) Daily Times*, 3 March 1893, 4; "Rev. Buswell's Trial," *Beatrice (Nebr.) Daily Express*, 28 February 1893, 4; "Statutes vs. Scriptures," *Beatrice (Nebr.) Daily Express*, 1; "Healing by Faith," *Beatrice (Nebr.) Daily Express*, 1 March 1893, 1; "The Buswell Trial," *Beatrice (Nebr.) Daily Express*, 2 March 1893, 1; "He Is Not Guilty," *Beatrice (Nebr.) Daily Express*, 3 March 1893, 1; "The Buswell Case," *Beatrice (Nebr.) Daily Express*, 4 March 1893, 4; and "Mockery—Disgrace," *Beatrice (Nebr.) Daily Express*, 6 March 1893, 4.

35. See "Rev. Buswell on Trial," *Beatrice (Nebr.) Daily Times*, 28 February 1893, 4, and "In the Jury's Hands," *Beatrice (Nebr.) Daily Times*, 2 March 1893, 4.

36. "In the Jury's Hands," *Beatrice (Nebr.) Daily Times*, 2 March 1893, 4. The *Beatrice Daily Express* echoed these sentiments, which seemed to reflect the community's anxiety about Christian Science *healing* but its desire to tolerate religious beliefs. See "He Is Not Guilty," *Beatrice (Nebr.) Daily Express*, 3 March 1893, 1.

37. State v. Buswell, 58 N.W. 728 (1894).

38. See the *Christian Science Journal* 11 (1893): 65–86, 91–94; 11 (July 1893): 165–70; and 12 (July 1894): 168–75.

39. *JAMA* welcomed the decision as well. See "'Christian Science,'" *JAMA* 22 (1894): 685.

40. J. V. Beghtol, "The Medical Law of Nebraska and Its Enforcement," *Proceedings of the Nebraska State Medical Society*, 1894, 19.

41. B. F. Crummer, "Report of Committee on Proposed Modification of the Medical Law of Nebraska," *Proceedings of the Nebraska State Medical Society*, 1894, 9.

42. Case file, State of Rhode Island, Gardner T. Swarts, Complainant, v. David Anthony, Respondent, case no. 18425, District Court of the Sixth Judicial District of the State of Rhode Island and Providence Plantations, 14–15, Phillips Memorial Library Archives, Providence College, Providence, Rhode Island.

43. Case file, State of Rhode Island, Gardner T. Swarts, Complainant, v. Walter E. Mylod, Respondent, case 18426. District Court of the Sixth Judicial District of the State of Rhode Island and Providence Plantations. Phillips Memorial Library Archives, Providence College, Providence, Rhode Island; State v. Mylod, 20 R.I. 632, 40 Atl. 753 (1898).

44. "The Board of Health of the State of Rhode Island . . . ," *JAMA* 32 (1899): 675.

45. "Editor's Table," *Christian Science Journal* 16 (1898): 448.

46. See In re First Church of Christ, Scientist, 55 Atl. 536 (1903), and *Christian Science Journal* 15 (1898): 717–30.

47. I have quoted from a revised version of this article that appeared as "To the Christian World" in *Christian Science Journal* 16 (1899): 665.

48. [Septimus J. Hanna], "Editor's Table," *Christian Science Journal* 16 (1898): 73. See also his "Editor's Table: Religious Liberty," *Christian Science Journal* 19 (1901): 133–40.

49. [Septimus J. Hanna], "Editor's Table," *Christian Science Journal* 15 (1898): 798.

50. See [Archibald McLellan], "Editor's Table," *Christian Science Journal* 20 (1903): 773–74; Archibald McLellan, "'An Invasion of Personal Liberties,'" *Christian Science Sentinel* 12 (2 April 1910): 610. Herbert W. Packard defended Christian Science healing under the terms of the Fourteenth Amendment's limitations on the state's police power in his "Christian Science and the Constitution," *Christian Science Journal* 23 (1905): 342–51. See also Clifford P. Smith, "Christian Science and Legislation," *Christian Science Journal* 23 (1905): 405–12, and Clarence A. Buskirk, "Medicine and the State," *Christian Science Sentinel* 13 (22 July

1911): 923–24. For the development of the First Amendment defense, see Willard S. Mattox, "Christian Science and Legislation," *Christian Science Journal* 22 (1904): 493–501; Clifford P. Smith, "Christian Science and Legislation: An Approach to the Subject," *Christian Science Journal* 23 (1905): 290–93; Alfred Farlow, *The Relation of Government to the Practice of Christian Science* (Boston, 1908; reprinted and revised from the *Government Magazine*); and Ernest C. Moses, "Religious Liberty in the United States," *Christian Science Journal* 35 (1918): 549–52.

51. "Christian Scientists Get Verdict," *Milwaukee Journal*, 15 April 1901, 5; A. C. Umbreit, ed., *Report of the Trial of Crecentia Arries and Emma Nichols, Charged with Practicing Medicine without a License, Before the Police Court of the City of Milwaukee, Giving a Verbatim Report of All the Testimony, Rulings of the Court, Arguments of Counsel and Decisions of the Presiding Judge* (Milwaukee: Edw. Keogh Press, 1900), 67. See also Elizabeth Barnaby Keeney, Susan Eyrich Lederer, and Edmond P. Minihan, "Sectarians and Scientists: Alternatives to Orthodox Medicine," in *Wisconsin Medicine: Historical Perspectives*, ed. Ronald L. Numbers and Judith Walzer Leavitt (Madison: University of Wisconsin Press, 1981), 60–61.

52. See Evans v. State, 9 Ohio Decisions 222 (1898). See the memorandum signed by Judge C. B. Elliott and attached to the demurrers granted by him for Mary Brookins (indicted as Mary Brookings) and Albert P. Meyer, cases 6123 and 6124, District Court for the Fourth Judicial District, County of Hennepin, Minnesota, Clerk's Office, District Court for the Fourth Judicial District, Minneapolis, Minnesota.

53. Case file, State of Ohio v. Oliver W. Marble, no. 9002, January Term, 1904, Supreme Court of the State of Ohio, Columbus, Ohio; State v. Marble, 73 N.E. 1063 (1905). See Reynolds v. United States, 98 U.S. 145 (1878).

54. For excellent overviews and analyses of relevant cases up to the time of their publication, see Irving E. Campbell, "Christian Science and the Law," *Virginia Law Register* 10 (1904): 285–300; John C. Myers, "Christian Science and the Law," *Law Notes* 12 (1908–9): 5–6; and Peter V. Ross, "Metaphysical Treatment of Disease as the Practice of Medicine," *Yale Law Journal* 24 (1914–15): 391–411.

55. "Christian Scientists and the Practice of Medicine," *Law Notes* 2 (1898): 142. For this view see also H. Gerald Chapin, "Christian Science and the Law," *Medico-Legal Journal* 17 (1899): 192–95.

56. "The position of LAW NOTES upon the standing of Christian Scientists . . . ," *Law Notes* 2 (1899): 181. See also Moritz Ellinger, "Christian Science and the Law," *Medico-Legal Journal* 17 (1899): 175–80.

57. William A. Purrington, "Manslaughter, Christian Science, and the Law," *Medical Record* 54 (1898): 761.

58. W. A. Purrington, "Eddyism before the Law," *Albany Law Journal* 62 (1900–1901): 358–62.

59. Carol Norton, "Christian Science and the Law," *Medico-Legal Journal* 17 (1899): 188–89.

60. Ibid., 190. John Carroll Lathrop seconded the views of Norton in his "Christian Science and the Law," *Medico-Legal Journal* 17 (1899): 196–97.

61. Starr, *Transformation of American Medicine,* 80. For further discussions of the formation of the professions in America, see Haber, *Quest for Authority and Honor,* and Nathan O. Hatch, ed., *The Professions in American History* (South Bend, Ind.: University of Notre Dame Press, 1988).

62. A. F. [Alfred Farlow], "Questions and Answers," *Christian Science Journal* 8 (1891): 497; A. F. [Alfred Farlow], "Questions and Answers," *Christian Science Journal* 10 (1892): 413.

63. Case file, State of Rhode Island, Gardner T. Swarts, Complainant, v. David Anthony, Respondent, 14–15.

64. Mary Baker Eddy, *Church Manual,* 11th ed. (Boston: Christian Science Publishing Society, 1899), 56–57.

65. "Notice," *Christian Science Journal* 7 (1889): 100–101.

66. E. J. Foster Eddy, "Address of Dr. E. J. Foster Eddy Before the N.C.S. Association, June 27, 1890," *Christian Science Journal* 8 (1890): 144. On the rise and decline of the urban dispensary movement in America, see Charles E. Rosenberg, "Social Class and Medical Care in Nineteenth-Century America: The Rise and Fall of the Dispensary," in *Sickness and Health in America: Readings in the History of Medicine and Public Health,* 3d ed. rev., ed. Judith Walzer Leavitt and Ronald L. Numbers (Madison: University of Wisconsin Press, 1997), 309–22, quotation on 318.

67. Caroline E. Linnell, "The Christian Science Reading Room," *Christian Science Sentinel* 11 (1909): 885–86.

68. See "'Christian Science' and the Practice of Medicine," *Western Medical Review* 3 (1898): 461.

69. "A Test for 'Christian Scientists,'" *JAMA* 34 (1900): 759.

70. A. S. Burdick, "Some Phases of Medical Practice Legislation," *Illinois Medical Journal* 3 (1901): 63. See also a letter from Dr. N. C. Steele of Chattanooga, Tennessee, in Nelson W. Wilson, "Illegal Medicine," *Buffalo Medical Journal* 39 (1899–1900): 284.

71. Ralph Elmergreen, "The Medical Mountebank," *Transactions of the State Medical Society of Wisconsin* 36 (1902): 190–91. See also H. Winnett Orr, "Medical Legislation in Nebraska," *Proceedings of the Nebraska State Medical Society,* 1900, 90–92; letter from Scott P. Child, M.D., of Kansas City in "Sacrificed to Eddyism," *JAMA* 41 (1903): 322–23; and R. Willman, "The Errors of Mind Healing," *Medical Herald* 29 (1910): 289.

72. On the popularization of science and health in America, see John C. Burnham, *How Superstition Won and Science Lost: Popularizing Science and Health in the United States* (New Brunswick, N.J.: Rutgers University Press, 1987), esp. chap. 2, pp. 45–84.

73. M., "Editor's Table," *Christian Science Journal* 20 (1903): 709.

74. Lee Zeunert Johnson, "Christian Science Committee on Publication: A Study of Group and Press Interaction" (Ph.D. diss., Syracuse University, 1963), 97–107. Regarding the exclusion of women from the Committees on Publication, see "Church By-Laws," *Christian Science Journal* 18 (1900): 35.

75. Mary Baker G. Eddy, "By-Laws," *Christian Science Journal* 17 (1900): 702.

76. "'Christian Science' and the Material Press Agent," *JAMA* 74 (1920): 1581.

77. For an overview of some of these activities, see Harold Walter Eickhoff, "The Organization and Regulation of Medicine in Missouri, 1883–1901" (Ph.D. diss., University of Missouri, 1964).

78. U. S. Wright, "President's Address," *Transactions of the Medical Association of the State of Missouri*, 1901, 24–25.

79. A. S. Burdick, "Some Phases of Medical Practice Legislation," *Illinois Medical Journal* 3 (1901): 61.

80. *Laws of Missouri, Forty-first General Assembly*, 2 January 1901 (Jefferson City, Mo.: Tribune Printing Company, 1901), 207, 208.

81. Wright, "President's Address," 23.

82. "Doctors Get First Blood," *Kansas City Star*, 16 February 1901, 10.

83. "'Science' in the House," *Kansas City Star*, 6 February 1901, 10.

84. *Journal of the House of Representatives of the Forty-first General Assembly of the State of Missouri, 1901* (Jefferson City, Mo.: Tribune Printing Company, 1901), 362–64; *Journal of the Senate of the Forty-first General Assembly of the State of Missouri, 1901* (Jefferson City, Mo.: Tribune Printing Company, 1901), 492–95.

85. "Doctors Get First Blood," *Kansas City Star*, 16 February 1901, 10.

86. Margery Fox, "Conflict to Coexistence: Christian Science and Medicine," *Medical Anthropology* 8 (1984): 295–96.

87. Edward W. Dickey, "Christian Science and Religious Liberty," *Michigan Law Review* 4 (1905–6): 266.

88. Oklahoma (1899): "Christian Science," *JAMA* 32 (1899): 563; Colorado (1903): Edward W. Dickey, "Christian Science and Religious Liberty," *Michigan Law Review* 4 (1905–6): 264; Nebraska (1905): ibid., 262.

89. Virginia (1903): "Eddyites and Osteopaths and Faith-Healers Defeated," *Sanitarian* 50 (1903): 435; Maryland (1908): "Faith Healing in Maryland," *JAMA* 50 (1908): 1448; Ohio (1915): "Christian Scientists and Death Certificates," *JAMA* 65 (1915): 2026.

90. Illinois (1898): W. [E. W. Weis], "Faith Healing," *Illinois Medical Journal* 1, n.s. (1899): 130; twenty-eight states by 1917: "Prayers for Profit," *JAMA* 68 (1917): 1701.

91. People v. Cole, 113 N.E. 790.

92. A. T. B. [Algernon Thomas Bristow], "Christian Science and the Law," *New York State Journal of Medicine* 10 (1910): 174.

93. "Legislative Notes," *New York State Journal of Medicine* 11 (1911): 144–49.

94. People v. Cole, 113 N.E. 790.

95. "Christian Science Measure Is Vetoed," *New York Times*, 25 April 1914, 11. See also "Healing Cults Seek Shelter of Laws," *New York Times*, 27 March 1914, 3; "The Healing Cults," *New York Times*, 4 April 1914, 14; [Editorial], "Dangerous Measures," *New York Times*, 5 April 1914, 14; [Editorial], "These Bills Must Be Vetoed," *New York Times*, 6 April 1914, 8; "Two X-Science Spe-

cials," *New York Times*, 12 April 1914, sect. 4, p. 4; "2,000 'Scientists' at Albany Hearing," *New York Times*, 14 April 1914, 2.

96. Farlow, *Relation of Government*, 6.

97. Carol Norton, "Working for the Cause," *Christian Science Journal* 21 (1903): 204.

Chapter Seven. Public Health and the Protection of Children

Epigraphs: L. W. [Watkins], "Christian Science," *Eclectic Medical Journal* 63 (1903): 343; Albert F. Gilmore, "Justice and the Law," *Law Notes* 24 (1920–21): 232.

1. " 'Christian Science' and Medical Practitioners," *JAMA* 33 (1899): 1049.

2. Judith Walzer Leavitt, *The Healthiest City: Milwaukee and the Politics of Health Reform* (Princeton: Princeton University Press, 1982), 3.

3. On the relation between filth, garbage, and disease, see Martin V. Melosi, *Garbage in the Cities: Refuse, Reform, and the Environment, 1880–1980* (College Station: Texas A&M University Press, 1981).

4. Regarding personal hygiene and public health, see Marilyn Thornton Williams, *Washing "the Great Unwashed": Public Baths in Urban America, 1840–1920* (Columbus: Ohio State University Press, 1991).

5. James B. Taylor, *Christian Science Considered* (Cincinnati: Cranston and Stowe, 1888), 41.

6. Clifford P. Smith, "Legal Status of Christian Science," *Christian Science Journal* 32 (1914): 327.

7. On the interplay between hygiene, germs, and public health, see Nancy Tomes, *The Gospel of Germs: Men, Women, and the Microbe in American Life* (Cambridge: Harvard University Press, 1998).

8. Case file, City of Davenport v. C. A. Owen, case 1548, Clerk's Office, District Court, Scott County, Davenport, Iowa.

9. "Owen Jerked Up," *Davenport (Iowa) Sunday Democrat*, 2 June 1895, 1.

10. "Board of Health," *Davenport (Iowa) Democrat*, 5 June 1895, 2.

11. "Owen's Trial," *Davenport (Iowa) Democrat*, 6 June 1895, 1.

12. "Ignorance No Crime," *Davenport (Iowa) Democrat*, 4 December 1895, 1; "Our Guardians," *Davenport (Iowa) Democrat*, 9 June 1895, 1.

13. For the antebellum roots of these efforts, see James S. Cassedy, *American Medicine and Statistical Thinking, 1800–1860* (Cambridge: Harvard University Press, 1984).

14. See Barbara Gutmann Rosenkrantz, *Public Health and the State: Changing Views in Massachusetts, 1842–1936* (Cambridge: Harvard University Press, 1972), 110–12.

15. Barbara Gutmann Rosenkrantz, "Cart before Horse: Theory, Practice and Professional Image in American Public Health, 1870–1920," *Journal of the History of Medicine* 29 (1974): 66–70.

16. A. S. Burdick, "Some Phases of Medical Practice Legislation," *Illinois*

Medical Journal 3 (1901): 60, 63. For further evidence of this resistance by physicians, see Thomas N. Bonner, *Medicine in Chicago, 1850–1950* (Madison, Wis.: American History Research Center, 1957), 206–9.

17. *The Christian Science Case, Prepared by Judge S. J. Hanna, Editor of the Christian Science Journal* (Boston: National Constitutional Liberty League, [1893?]), 1.

18. Richard Olding Beard, "The Relation of the Medical Profession to the Enactment of Legislation upon Questions of Public Health," *Transactions of the Minnesota State Medical Association*, 1903, 307.

19. "Drugs, Hygiene and Hypnotism," *Christian Science Journal* 9 (1892): 460.

20. Ibid.

21. [Septimus J. Hanna], "Editor's Table," *Christian Science Journal* 15 (1898): 797.

22. "Editor's Table," *Christian Science Journal* 15 (1897): 132.

23. For descriptions of this later debate, see Ronald L. Numbers, *The Creationists* (New York: Alfred A. Knopf, 1992), 37–53, 336–37, and George M. Marsden, *Fundamentalism and American Culture: The Shaping of Twentieth-Century Evangelicalism, 1870–1925* (New York: Oxford University Press, 1980), 212–21.

24. "Scientists Disobey Law," *Kansas City (Mo.) Star*, 4 November 1897, 1.

25. Ibid.

26. "Mrs. Baird Is Arrested," *Kansas City (Mo.) Star*, 5 November 1897, 1.

27. [Editorial], "Settle It Now," *Kansas City (Mo.) Star*, 5 November 1897, 4.

28. For the history of Christian Science in Kansas City, see Jessie Bain Cooper, "Christian Science," in *Encyclopedia of the History of Missouri*, ed. Howard L. Conard (New York: Southern History Company, 1901).

29. "Mrs. Baird Is Arrested," *Kansas City (Mo.) Star*, 5 November 1897, 1.

30. "Not All Are with Mrs. Baird," *Kansas City (Mo.) Star*, 8 November 1897, 1.

31. Carol Norton, "Christian Science and the Law," *Medico-Legal Journal* 17 (1899): 189.

32. Alfred Farlow related these experiences in a letter to the newspaper. See "Divine Power as a Disinfectant," *Kansas City (Mo.) Star*, 7 November 1897, 1.

33. "May Stand by Mrs. Baird," *Kansas City (Mo.) Star*, 7 November 1897, 1.

34. Baird was convicted in the city court (1897) and then reconvicted in the Jackson County criminal court (1898) before the reversal by the state court of appeals (1902). See Kansas City v. Amanda J. Baird, 92 Missouri Appeal Reports 204.

35. "'Christian Science' and the Maryland Law," *JAMA* 34 (1900): 626–27; W. A. Purrington, "Eddyism before the Law," *Albany Law Journal* 62 (1900–1901): 358–62; Robert T. Morris, "Address before the New York State Assembly Committee on Public Health," *JAMA* 36 (1901): 567–69; and "The Ministers' and Healers' Bill," *Journal-Lancet* 35 (1915): 156–57.

36. This reportedly occurred in Anderson, Indiana. See "'Died from Sin and Fear,'" *New York Times*, 13 November 1895, 1.

37. For an example of a converted physician signing death certificates for

Scientists, see the case of New York's Dr. Charles Griffin Pease in "Faith Healers' Methods," *New York Times*, 20 December 1900, 6. Pease subsequently faced trial for failing to report the contagious disease of the person for whom he had signed the death certificate.

38. See "'Christian Science' and Death Certificates," *JAMA* 32 (1899): 1185; "'Christian Science' and Death Certificates," *JAMA* 33 (1899): 1657; "Christian Science a Menace to the Public Weal," *Boston Medical and Surgical Journal* 142 (1900): 283; "Disease and Deaths among Christian Scientists," *Northwestern Lancet* 22 (1902): 94; "The Making of Death Certificates for Eddyite Victims," *JAMA* 48 (1907): 1780; "Death Certificates by Unlicensed Healers," *JAMA* 66 (1916): 50; "Christian Science; The Greatest Delusion of the Age," *Journal of the Indiana State Medical Association* 14 (1921): 270; and "Christian Scientists Pretend by Some Sort of Mental Process That They Can Cure the Most Virulent Communicable and Oftentimes Incurable Diseases," *Illinois Medical Journal* 39 (1921): 175.

39. I have quoted from "Obey the Law," *Christian Science Journal* 18 (1901): 724. The same message appeared in the *Boston Herald* on 17 February 1901 and elsewhere as well.

40. Spoken in an interview published in the *New York Herald* on 1 May 1901 and quoted here as published in "Mrs. Eddy Talks," *Christian Science Journal* 19 (1901): 144–45.

41. On the nature and treatment of diphtheria, see Anne Hardy, "Tracheotomy versus Intubation: Surgical Intervention in Diphtheria in Europe and the United States, 1825–1930," *Bulletin of the History of Medicine* 66 (1992): 536–59, and Charles R. King, *Children's Health in America: A History* (New York: Twayne, 1993), 76–79. Regarding the varied rates of decline in diphtheria death rates, see Gretchen A. Condran, Henry Williams, and Rose A. Cheney, "The Decline in Mortality in Philadelphia from 1870 to 1930: The Role of Municipal Services," in *Sickness and Health in America: Readings in the History of Medicine and Public Health*, 3d ed. rev., ed. Judith Walzer Leavitt and Ronald L. Numbers (Madison: University of Wisconsin Press, 1997), 452–66. On physicians' growing support of antitoxin, see Rosenkrantz, "Cart before Horse," 69–72.

42. "Christian 'Science,'" *JAMA* 25 (1895): 591.

43. For these and other details of this case, see Rex v. Lewis, 6 Ont. L. R. 132, 1 B.R.C. 732 (1903), 7 Can. Crim. Cases 261, Anno. 10 ALR 1150; "Lawyer Startles the Court," *Toronto World*, 2 November 1901, 9; "Evidence in Lewis Case," *Toronto World*, 4 November 1901, 6; "The Manslaughter Case," *Toronto World*, 5 November 1901, 5; "Lewis Was Found Guilty," *Toronto World*, 6 November 1901, 7; [Editorial], "The Christian Science Case," *Toronto World*, 7 November 1901, 4; and "Judgment Day at the Hall," *Toronto World*, 30 June 1903, 4. On the reaction of Canadian physicians, see James H. Richardson, "The Recent Christian Science Trial," *Canadian Journal of Medicine and Surgery* 10 (1901): 403–9. For an overview of the relationships between Ontario physicians and alternative healers, including Christian Scientists, see Ronald Hamowy, *Canadian Medicine: A Study in Restricted Entry* (Toronto: Fraser Institute, 1984), 184–94.

44. "Christian Scientists Must Stand Trial," *New York Times*, 31 October 1902, 1. For a Christian Science father's recollections of one of the cases (1897 or 1898) that the Westchester County grand jury failed to indict, see a memorandum by John M. Goodwin dated 3 January 1909 in Alfred Farlow Vault box 28, Archives of the Mother Church, Christian Science Center, Boston.

45. "Healers Flock to Lathrop's Defence," *New York Evening Journal*, 24 October 1902, 5.

46. "Use Your Judgment Says 'Mother' Eddy," *New York Evening Journal*, 26 November 1902, 5. This advice appeared as well under Eddy's name as "Wherefore?" *Christian Science Journal* 20 (1902): 578–79.

47. "Christian Scientists, Vaccination and Contagion," *Northwestern Lancet* 21 (1901): 129.

48. The *New York Medical Journal* spoke of "clear sacrifice." See "An Editorial Symposium on Mrs. Eddy's Latest Pronunciamento," *Boston Medical and Surgical Journal* 147 (1902): 604, which excerpted reactions from several other medical journals as well.

49. "Christian Science's Latest Bulletin," *Northwestern Lancet* 23 (1903): 11.

50. "How Prayer Failed to Cure Diphtheria," *New York Times*, 23 October 1902, 16.

51. "The Surrender of Eddyism," *JAMA* 39 (1902): 1326.

52. Among a variety of possible explanations for this growing concern about children, scholars have suggested changes in fertility, challenges to Calvinism, an altering definition of family, the expansion of individualism, and new psychological definitions of childhood. For example, see Carl N. Degler, *At Odds: Women and the Family in America from the Revolution to the Present* (New York: Oxford University Press, 1980), and Joseph F. Kett, *Rites of Passage: Adolescence in America, 1790 to the Present* (New York: Basic Books, 1977).

53. See Mary Ann Mason, *From Father's Property to Children's Rights: The History of Child Custody in the United States* (New York: Columbia University Press, 1994); Joseph M. Hawes, *The Children's Rights Movement: A History of Advocacy and Protection* (Boston: Twayne, 1991), esp. chaps. 2–3; Walter I. Trattner, *From Poor Law to Welfare State: A History of Social Welfare in America*, 2d ed. (New York: Free Press, 1979), esp. chaps. 5–6; and Robert H. Bremner, ed., *Children and Youth in America: A Documentary History*, vol. 2, *1866–1932* (Cambridge: Harvard University Press, 1971).

54. See J. C. Bierwirth, "The Relations of the So-Called 'Christian Science Cure' and 'Faith Cure' to Public Health and Boards of Health, and to the Medical Profession," *New York Medical Journal* 50 (1889): 713–16. Bierwirth was the physician to the Brooklyn Society for the Prevention of Cruelty to Children. Although the two specific cases discussed in this speech to the Kings County Medical Association involved members of the New Evangelists, he applied his conclusions to all religious healers, including Christian Scientists.

55. The Child Health Organization and the American Child Hygiene Association joined in 1922 to become the American Child Health Association. See

King, *Children's Health in America*, 127–29, and Roger Cooter, ed., *In the Name of the Child: Health and Welfare, 1880–1940* (New York: Routledge, 1992).

56. On the origins of pediatrics as an American medical specialty, see Sydney A. Halpern, *American Pediatrics: The Social Dynamics of Professionalism, 1880–1980* (Berkeley: University of California Press, 1988), esp. chaps. 3–5, pp. 35–109.

57. Michael Grossberg, *Governing the Hearth: Law and the Family in Nineteenth-Century America* (Chapel Hill: University of North Carolina Press, 1985), 279–80.

58. Queen v. Senior, 1 Q. B. 283 (1899), 19 Cox C. C. 219. For discussion of this case and its legal contexts, see Irving E. Campbell, "Christian Science and the Law," *Virginia Law Register* 10 (1904): 293–95; I. H. Rubenstein, "Faith Healing, Legal Aspects," in *A Treatise on Contemporary Religious Jurisprudence* (Chicago: Waldain Press, 1948), 49–51. For an overview of the legal difficulties of Christian Scientists in Great Britain, see Claire F. Gartrell-Mills, "Christian Science: An American Religion in Britain, 1895–1940" (Ph.D. diss., Wolfson College, Oxford University, 1991), esp. chap. 7, "Christian Science and the Law," 210–28.

59. Rex v. Lewis, 6 Ont. L. R. 132 (1903).

60. "Lewis Was Found Guilty," *Toronto World*, 6 November 1901, 7.

61. "Child Died without Medical Attendance," *New York Times*, 22 October 1902, 16. On the early history of Christian Science in New York City, see Sarah Gardner Cunningham, "A New Order: Augusta Emma Simmons Stetson and the Origins of Christian Science in New York City, 1886–1910" (Ph.D. diss., Union Theological Seminary, 1994).

62. "Coroner Is after Science Healers," *New York Evening Journal*, 22 October 1902, 4.

63. For Lathrop's view, see "Child Died without Medical Attendance," *New York Times*, 22 October 1902, 16. On McCracken's view, see "Ready to Aid 'Healer,'" *New York Times*, 25 October 1902, 6.

64. "Healers Flock to Lathrop's Defence," *New York Evening Journal*, 24 October 1902, 5.

65. "Healers Call Mass Meeting for Quimbys," *New York Evening Journal*, 25 October 1902, 4. Newspaper estimates of the size of the legal defense fund ranged from $10,000 to $100,000. See "Quimby Back to Work, Clings to Faith Cure," *New York Evening Journal*, 24 October 1902, 5; "Churchmen on Science Jury," *New York Evening Journal*, 28 October 1902, 5; and "'Not Guilty,' Plead Quimbys," *New York Evening Journal*, 6 November 1902, 6.

66. On the anxieties of parents, see "Coroner Is after Science Healers," *New York Evening Journal*, 22 October 1902, 4; "Healers Call Mass Meeting for Quimbys," *New York Evening Journal*, 25 October 1902, 4. For anti–Christian Science sermons delivered at the time of the inquest, see "Dr. Parkhur[s]t Says Mrs. Eddy Is a Liar," *New York Evening Journal*, 27 October 1902, 3, and "Attacks

Christian Science," *New York Times*, 27 October 1902, 14. For the religious affiliations of the jurymen, see "Churchmen on Science Jury," *New York Evening Journal*, 28 October 1902, 5.

67. [Editorial], "The Murder of Children by Christian Scientists," *New York Evening Journal*, 27 October 1902, 12.

68. "Christian Science Case," *New York Times*, 26 November 1902, 1.

69. "Scientist Suggests Doctors Be Indicted," *New York Evening Journal*, 30 October 1902, 8.

70. See also Alfred Farlow, *The Relation of Government to the Practice of Christian Science* (Boston, 1908), 19–22; reprinted and revised from *Government Magazine*).

71. On Christian Science support, see "Criminal Faith-Healing," *New York Times*, 14 October 1903, 1. On the grounds of appeal, see "Faith Curist's Belief," *New York Times*, 21 November 1902, 2.

72. *Black's Law Dictionary* defines *parens patriae* as "literally 'parent of the country,'" referring "traditionally to [the] role of [the] state as sovereign and guardian of persons under legal disability."

73. [Editorial], "Murder Declared Unlawful," *New York Times*, 15 October 1903, 8. Regarding physicians' and lawyers' reaction to the Pierson decision, see "Faith Healing and Criminal Neglect in New York," *JAMA* 41 (1903): 1028–29; "The Legal Status of Faith Healing," *Boston Medical and Surgical Journal* 149 (1903): 467–68; "The Case of People v. Pierson . . .," *Law Notes* 7 (1903): 21–22; and "'A GOOD Deal of Jolt in the Jaw' . . .," *Law Notes* 7 (1903): 163.

74. Pennsylvania: Commonwealth v. Hoffman, 29 Penn. Co. Ct. R. 65 (1903); Indiana: State v. Chenoweth, 163 Ind. 94, 71 N. E. 197 (1904); Maine: State v. Sandford, 99 Me. 441, 59 Atl. 597 (1905), reversed on a technicality; and Georgia: Bennett v. Ware, 4 Ga. A. 293, 61 S. E. 546 (1908); Oklahoma: 6 Okla. Cr. 110, 116 Pac. 345 (1911).

75. "Christian Scientists Won a Verdict," *Los Angeles Daily Times*, 30 November 1902, sec. 4, p. 8.

76. "Christian Scientists Freed," *New York Times*, 8 August 1905, 7.

77. On Byrne's case, see "Jury Would Make Eddyism Felony," *New York Times*, 18 May 1907, 1, and "Jail Sentence for Eddyite," *New York Times*, 3 August 1907, 14.

78. C. C. Cawley, "Criminal Liability in Faith Healing," *Minnesota Law Review* 39 (1954): 57.

79. [Mary Baker Eddy], "Christian Science and China," *Christian Science Journal* 24 (1906): 49. On 2 March 1906 Eddy wrote this advice to Sarah Pike Conger, C.S.B., a missionary in China. Although Eddy was warning Conger about the dangers in China, she clearly drew upon her current difficulties in America to make the point.

80. Archibald McLellan, "Christian Science and Legislation," *Christian Science Sentinel* 8 (27 January 1906): 344.

81. Mary Baker Eddy, *Church Manual*, 66th ed. (Boston: Christian Science Publishing Society, 1906; 1907 printing), 102.

Chapter Eight. Century of Promise, Then Peril

Epigraphs: Committee on Bioethics, American Academy of Pediatrics, "Religious Exemptions from Child Abuse Statutes," *Pediatrics* 81 (1988): 170–71; "Speaking for Children," *Christian Science Monitor*, 17 August 1989, as quoted in *Freedom and Responsibility: Christian Science Healing for Children* (Boston: First Church of Christ, Scientist, 1989), 91.

1. According to Eddy's death certificate, she died of "natural causes, probably pneumonia." See death certificate of Mary (Baker) Eddy, date of death 3 December 1910, date of registry 9 December 1910, no. 461, libra 6, folio 104, City of Newton, Commonwealth of Massachusetts.

2. For insights into this process of adapting religious movements to a dead prophet, see Timothy Miller, *When Prophets Die: The Postcharismatic Fate of New Religious Movements* (Albany: State University of New York Press, 1991).

3. Charles S. Braden, *Christian Science Today: Power, Policy, Practice* (Dallas: Southern Methodist University Press, 1958), 61–95.

4. For evidence of this decline, see Olga M. Chaffee, "Report of the Clerk of the Mother Church," *Christian Science Journal* 110 (1992): 26–28.

5. *A Century of Christian Science Healing* (Boston: Christian Science Publishing Society, 1966), 239.

6. Quotation from Robert Peel, *Mary Baker Eddy* (New York: Holt, Rinehart, Winston, 1966–77), 3:239.

7. Arthur Edmund Nudelman, "Christian Science and Secular Medicine" (Ph.D. diss., University of Wisconsin, 1970), quotation on 138–39. See also Arthur E. Nudelman and B. E. Nudelman, "Health and Illness Behavior of Christian Scientists," *Social Science and Medicine* 6 (1972): 253–62, and Arthur E. Nudelman, "Use of a Medical Facility by Christian Scientists," *Southern Medical Journal* 79 (1986): 437–40.

8. Stephen Gottschalk, "Spiritual Healing on Trial: A Christian Scientist Reports," *Christian Century* 105 (1988): 604.

9. Robert Peel, "Christian Science and Death," unpublished manuscript in author's possession, 2. A version of this paper appeared in *Religion and Bereavement*, ed. Austin H. Kutscher and Lillian G. Kutscher (New York: Health Sciences, 1972). See also Lucy Jayne Kamua, "Systems of Belief and Ritual in Christian Science" (Ph.D. diss., University of Chicago, 1971), 116–18.

10. *Century of Christian Science Healing*, 254.

11. For example, see Stephen Gottschalk, *The Emergence of Christian Science in American Religious Life* (Berkeley: University of California Press, 1973), and *Century of Christian Science Healing*.

12. Alan A. Aylwin, "Help for the Addict," *Christian Science Journal* 89 (1971): 539.

13. Nathan A. Talbot, "The Question of Drinking," *Christian Science Sentinel* 84 (1982): 1742–46. See also DeWitt John, *The Christian Science Way of Life*, with *A Christian Scientist's Life* by Erwin D. Canham (Englewood Cliffs, N.J.: Prentice-Hall, 1962).

14. Gale E. Wilson, "Christian Science and Longevity," *Journal of Forensic Sciences* 1 (1956): 43–60; William Franklin Simpson, "Comparative Longevity in a College Cohort of Christian Scientists," *JAMA* 262 (1989): 1657–58. Corrections of two numerical errors in the Simpson article appeared in *JAMA* 262 (1989): 3000; letters to the editor and a reply by Simpson appeared as "Comparative Longevity of Christian Scientists," *JAMA* 263 (1990): 1634.

15. Lee Zeunert Johnson, "Christian Science Committee on Publication: A Study of Group and Press Interaction" (Ph.D. diss., Syracuse University, 1963), 78. Regarding the charges, Scientist spokesperson Stephen Gottschalk reported $8 to $12; see his letter in "Spiritual Healing and the Law: A Dispute," *Christian Century* 105 (1988): 928. But the Christian Science church in its amicus curiae brief in the Laurie Walker case reported "seven to fifteen dollars per day of treatment"; see Walker v. Superior Court of Sacramento County, S.F. no. 24996, Supreme Court of California; 47 Cal. 3d 112; 763 P. 2d 852; 1988 Cal. LEXIS 252; 253 Cal. Rptr. 1.

16. *Christian Science Monitor* 77 (27–30 November 1984).

17. *Christian Science Monitor* 77 (27 November 1984): 14.

18. Lewis Hubner, "The Function of Our Sanatoriums," *Christian Science Journal* 92 (1974): 149–50.

19. "A Christian Science Sanatorium," *Journal-Lancet* 36 (1916): 607; "Eddian Science Singular Sanitarium 'Certain Cases Not Received,'" *California State Journal of Medicine* 18 (1920): 240.

20. Marco Frances Farley, "Nursing: A Truly Satisfying Career," *Christian Science Journal* 97 (1979): 666.

21. Robert Peel, *Health and Medicine in the Christian Science Tradition: Principle, Practice, and Challenge* (New York: Crossroad, 1988), 95, 141 n. 14; *Century of Christian Science Healing*, 242. Since 1997, accreditation of nursing facilities has been in the hands of the Commission for Accreditation of Christian Science Nursing Organizations/Facilities, Inc., an organization independent of direct church control. See "Latest Media Information," official home page of the Church of Christ, Scientist, www.tfccs.com/GV/TMC/media.html, accessed 11 July 2000. On 1 May 2000 the United States Court of Appeals for the Eighth Circuit upheld the constitutionality of Medicare and Medicaid reimbursement of non-medical nursing. Suit had been brought on this matter by CHILD, Inc. See "Latest Media Information," official home page of the Church of Christ, Scientist, www.tfccs.com/GV/TMC/media.html, accessed 11 July 2000.

22. "From the Directors: The Christian Science Pleasant View Home," *Christian Science Journal* 43 (1925): 509.

23. Board of Directors, "The Christian Science Pleasant View Home," *Christian Science Journal* 44 (1926): 490–91.

24. On the membership numbers, see Gordon W. Rice, "Pseudomedicine," *JAMA* 58 (1912): 362. John H. Thompson, C.S.B. (see chapter 3), circulated membership applications and solicited donations for this League, although he wanted the Christian Science presence in the League to remain discreet. See John H. Thompson to Mr. and Mrs. Abrams, 21 October 1910; John H. Thompson to Miss Bassett, 21 October 1910; and John H. Thompson to Mr. Jacob, 28 October 1910, John H. Thompson Papers, Duke University Library, Durham, North Carolina.

25. "A Birds-Eye View of the National League for Medical Freedom," *Medical Freedom* 4 (1915): 117. For further information on the League and its activities, see further issues of *Medical Freedom; First Report of the National League for Medical Freedom* (New York: National League for Medical Freedom, 1910), and Martin Kaufman, *Homeopathy in America: The Rise and Fall of a Medical Heresy* (Baltimore: Johns Hopkins University Press, 1971), 162–65. Regarding the involvement of Christian Scientists, see Augusta E. Stetson, *Should Christian Scientists Become Identified with Any Medical League?* (Author, 1911), and "Mrs. Eddy and the National League for Medical Freedom," *JAMA* 58 (1912): 559.

26. *Congressional Record*, 6 July 1911, 2659–75. For a discussion of Works's speech and the reactions to it see Patricia Spain Ward, "Simon Baruch: Rebel in the Ranks of Medicine, 1840–1921" (Ph.D. diss., University of Wisconsin—Madison, 1990), 440–43. For further evidence of Christian Science opposition to the national Department of Health, see Archibald McLellan, "Public Sentiment Awakened," *Christian Science Sentinel* 13 (22 July 1911): 930.

27. Arthur J. Viseltear, "Compulsory Health Insurance in California, 1915–18," *Journal of the History of Medicine* 24 (1969): 151–82, quotations on 177.

28. Ronald L. Numbers, *Almost Persuaded: American Physicians and Compulsory Health Insurance, 1912–1920* (Baltimore: Johns Hopkins University Press, 1978), quotation on 81.

29. "Ohio Finds a Better Way," *JAMA* 72 (1919): 497–98.

30. "The Result in Ohio," *JAMA* 72 (1919): 578.

31. "Ohio Finds a Better Way," *JAMA* 72 (1919): 671.

32. "The Numerical Strength of Christian Science," *Illinois Medical Journal* 38 (1920): 35.

33. Clifford P. Smith, "Legal Status of Christian Science," *Christian Science Journal* 32 (1914): 327.

34. Gilmore, "Justice and the Law," 231–33.

35. For this distinction between "exemption" and "exclusion," which Johnson applied to the second half of the twentieth century, see Johnson, "Christian Science Committee on Publication," 543–46. On opposition to fluoridation, see ibid., 555–62.

36. C. C. Cawley, "Criminal Liability in Faith Healing," *Minnesota Law Review* 39 (1954): 67; *Time*, 24 May 1954, 55; "Religious Exemptions to Child Neglect Laws Still Being Passed Despite Convictions of Parents," *JAMA* 264 (1990): 1226–29, 1233.

37. Thomas Novotny et al., "Measles Outbreaks in Religious Groups Exempt from Immunization Laws," *Public Health Reports* 103 (1988): 49–54; "Outbreak of Measles among Christian Science Students—Missouri and Illinois, 1994," *Morbidity and Mortality Weekly Report* 43 (1994): 463–65; "247 Cases of Measles in 10 States Are Traced to a Skier in Colorado," *Los Angeles Times*, 2 September 1994, A20; Rita Swan, "Children, Medicine, Religion, and the Law," *Advances in Pediatrics* 44 (1997): 491–543.

38. For excellent introductions to the legal sources and issues involving children and religious healing, see Elena M. Kondos, "The Law and Christian Science Healing for Children: A Pathfinder," *Legal Reference Services Quarterly* 12 (1992): 5–71, and Rita Swan, "Children, Medicine, Religion, and the Law," *Advances in Pediatrics* 44 (1997): 491–543.

39. On the 1918–19 influenza pandemic, see "Attitude of Christian Scientists in the Present Epidemic of Influenza," *JAMA* 71 (1918): 1337, 1603, 1766, and "Christian Scientists Pretend by Some Sort of Mental Process That They Can Cure the Most Virulent Communicable and Oftentimes Incurable Diseases," *Illinois Medical Journal* 39 (1921): 175. On the use of statistics, see "'Religious Daily' Reports to Falsehood," *JAMA* 62 (1914): 460–61; "Knowledge versus Superstition," *JAMA* 62 (1914): 463; "'Christian Science' and Sloppy Thinking," *JAMA* 74 (1920): 1460–61, 1593; and Albert F. Gilmore [New York City Christian Scientist], "Justice and the Law," *Law Notes* 24 (1920–21): 231–32.

40. "Protest from a Christian Scientist," *Law Notes* 7 (1904): 220. Farlow wrote this letter to protest the journal's use of the term Christian Scientist in conjunction with the Pierson case.

41. [Editorial], "Why Convictions Are So Infrequent," *New York Times*, 6 May 1920, 10.

42. "Report of Committee on Publication, Annual Meeting of the Mother Church, 1926," *Christian Science Journal* 44 (1926): 209.

43. Robert L. Trescher and Thomas N. O'Neill Jr., "Medical Care for Dependent Children: Manslaughter Liability of the Christian Scientist," *University of Pennsylvania Law Review* 109 (1960): 203–17.

44. For example, see Craig v. Maryland, 155 A. 2d 648 (Md., 1959); "Criminal Law—Manslaughter Conviction for Failure to Provide Medical Aid to Child because of Religious Belief Reversed," *De Paul Law Review* 9 (1960): 274.

45. On inclusion in health insurance, see I. H. Rubenstein, "Claim Cases by Christian Scientists," *Insurance Law Journal*, no. 477 (1962): 774–78; DeWitt John, "Recognition of Christian Science Treatment," *Insurance Law Journal*, no. 480 (1963): 18–22; "From Brief of Amicus Curiae on Behalf of the First Church of Christ, Scientist in Support of Petitioner Laurie Grouard Walker in the Supreme Court of the State of California," in *Freedom and Responsibility*, 61–64; and Matt Pommer, "Insurance for Prayer Is Sought," *Madison (Wis.) Capital Times*, 1 September 1988, 29, 34.

46. For the broader legal context of drugless healing during this period, see

John F. Hansler, "The Standard of Care of the Drugless Healer," *Washington Law Review* 27 (1952): 38–65, and I. H. Rubenstein, *A Treatise on Contemporary Religious Jurisprudence* (Chicago: Waldain Press, 1948).

47. Leo Damore grippingly told the story of Dorothy Sheridan and its aftermath in *The "CRIME" of Dorothy Sheridan* (New York: Arbor House, 1978), quotation on 312–13.

48. Stephen Goode, "When Doctors and Devotion Clash," *Insight*, 20 June 1988, 57.

49. C. C. Cawley, "Criminal Liability in Faith Healing," *Minnesota Law Review* 39 (1954): 49.

50. Prince v. Massachusetts, 321 U.S. 158; quotations on 167, 170.

51. Catherine W. Laughran, "Religious Beliefs and the Criminal Justice System: Some Problems of the Faith Healer," *Loyola of Los Angeles Law Review* 8 (1975): 411; Wisconsin v. Yoder, 406 U.S. 205 (1972).

52. Robert Peel, *Spiritual Healing in a Scientific Age* (San Francisco: Harper and Row, 1987), 114. These parents had been lifelong Christian Scientists who subsequently sued their practitioners and the Mother Church for "negligence and misrepresentation." See Brown v. Laitner, 435 N.W. 2d 1 (Mich. 1989).

53. For an account of Matthew Swan's last days, see Ramona Cass, "We Let Our Son Die: The Tragic Story of Rita and Doug Swan," *Journal of Christian Nursing* 4 (1987): 4–8.

54. The civil suit has come to be a mark of CHILD's efforts to halt the religious healing of children.

55. On the Swans and their activities, see Brown v. Laitner, 435 N.W. 2d 1 (Mich. 1989); Richard N. Ostling, "Matters of Faith and Death," *Time*, 16 April 1984, 42; Laura Sessions Stepp, "More Children's Deaths Laid to Parents' Faith," *Washington Post*, 9 January 1988, A3; Tamara Jones, "Child Deaths Put Faith on Trial," *Los Angeles Times*, 27 June 1989, sec. 1, p. 16; "Effort Seeks to Force Child Medical Care Regardless of Beliefs," *Los Angeles Times*, 27 November 1993, B4; and Fred Bayles, "Religion, Health Care Collide on Issue of Treating Children," *Riverside County (Calif.) Press-Enterprise*, 29 November 1993, A-1, A-10. By the end of 1993, Massachusetts, Hawaii, and South Dakota had revoked religious exemptions from child abuse laws, and Iowa, Minnesota, Missouri, Maryland, Delaware, and Kentucky legislatures had considered but failed to revoke such exemptions. See Bayles, "Religion, Health Care," A-1.

56. Diana Brahams, "Medicine and the Law," *Lancet* 336 (1990): 107. Between 1980 and 1993, "more than 40 criminal prosecutions in the United States have been brought against parents for the death of a child resulting from religiously motivated medical neglect." See Andrew A. Skolnick, "Christian Science Church Loses First Civil Suit on Wrongful Death of a Child," *JAMA* 270 (1993): 1781.

57. For press coverage of these trials, see "Parents, Religious Practitioner Charged after Tot Dies," *Orange County (Calif.) Register*, 23 June 1984; "Defense Attorney Cites Rapid Death of Ill Child," *Riverside County (Calif.) Press-Enterprise*, 14 March 1985, A-8; Herb Michelson, "Prayer Healing Faces Courtroom Chal-

lenges in California," *Riverside County (Calif.) Press-Enterprise*, 26 August 1986, A-8; Bob Egelko, "State High Court Agrees to Hear Faith-Healing Case," *Riverside County (Calif.) Press-Enterprise*, 28 March 1986, A-4; Kenneth Woodward, Sue Hutchison, and James B. Meadow, "The Graying of a Church: Christian Science's Ills," *Newsweek*, 3 August 1987, 60; Laura Sessions Stepp, "More Children's Deaths Laid to Parents' Faith," *Washington Post*, 9 January 1988, A3; Philip Hager, "Prosecutor Seeks Trial for Christian Scientist," *Los Angeles Times*, 9 March 1988, sec. 1, p. 23; Goode, "When Doctors and Devotion Clash," 56–57; Philip Hager, "Prosecution OKd in Prayer Case," *Los Angeles Times*, 11 November 1988, sec. 1, pp. 1, 24; Jones, "Child Deaths Put Faith on Trial," 1, 16; David G. Savage, "High Court Permits Prosecution in Faith-Healing Death," *Los Angeles Times*, 20 June 1989, sec. 1, p. 16; "Couple Cleared of Involuntary Manslaughter," *Riverside County (Calif.) Press-Enterprise*, 5 August 1989, A-3; Linda Chong, "Couple Who Failed to Get Treatment for Son Acquitted," *Los Angeles Times*, 18 February 1990, sec. B, pp. 1, 5; "Mother Guilty in Death of Girl," *Los Angeles Times*, 25 June 1990, A20; "Christian Scientists Convicted in Death," *Los Angeles Times*, 5 July 1990, A26; "Court Clears Couple in Child's Death," *Los Angeles Times*, 21 September 1991, A25; and "Court Overturns Conviction of Christian Science Couple," *Los Angeles Times*, 12 August 1993, A32. See also Richard J. Brenneman, "Nestling's Faltering Flight: The Short Life and Death of Seth Ian Glaser," in *Deadly Blessings: Faith Healing on Trial* (Buffalo, N.Y.: Prometheus Books, 1990).

58. For an outstanding discussion of the ways the trials of the 1980s raised in significant new ways the question whether the prosecution of religious healers violated the Fourteenth Amendment's guarantee of due process, see Janna C. Merrick, "Christian Science Healing of Minor Children: Spiritual Exemption Statutes, First Amendment Rights, and Fair Notice," *Issues in Law and Medicine* 10 (1994): 321–42. For details of the Walker case, see Walker v. Superior Court, Sacramento County, Sup. Ct. Cal., 1-5-89, 763 P2d 852; Michelson, "Prayer Healing"; Egelko, "State High Court"; Hager, "Prosecutor Seeks"; Hager, "Prosecution OKd"; Savage, "High Court Permits"; and "Mother Guilty."

59. For an excellent overview of the legal issues involved in these three 1984 California cases, see JoAnna A. Gekas, "California's Prayer Healing Dilemma," *Hastings Constitutional Law Quarterly* 14 (1986–87): 395–419.

60. For a useful critique of the supreme court's decision, see John T. Gathings Jr., "When Rights Clash: The Conflict between a Parent's Right to Free Exercise of Religion versus His Child's Right to Life," *Cumberland Law Review* 19 (1989): 585–616.

61. On non–Christian Science cases, see Jones, "Child Deaths Put Faith on Trial," 16.

62. C. Kempe et al., "The Battered Child Syndrome," *JAMA* 181 (1962): 17–24. My discussion in this paragraph is based on Norman Gevitz, "Christian Science Healing and the Health Care of Children," *Perspectives in Biology and Medicine* 34 (1991): 421–38, quotations on 429–30.

63. On the McMartin case, see Paul Eberle and Shirley Eberle, *The Abuse of*

Innocence: The McMartin Preschool Trial (New York: Prometheus Books, 1993). For claims of social hysteria and the fear of sex abuse, see Richard A. Gardner, *Sex Abuse Hysteria: Salem Witch Trials Revisited* (Cresskill, N.J.: Creative Therapeutics, 1990). For examples of media reflection on the child abuse cases of the 1980s and early 1990s, see Lawrence Wright, "Child-Care Demons," *New Yorker*, 3 October 1994, 5–6; Franklin E. Zimring, "Paranoia on the Playground," *Los Angeles Times*, 11 November 1996, B5; and Carol Tavris, "A Day-Care Witch Hunt Tests Justice in Massachusetts," *Los Angeles Times*, 11 April 1997, B9.

64. American Academy of Pediatrics, "Religious Exemptions," 169–71.

65. Goode, "When Doctors and Devotion Clash," 56.

66. *New York Times* 19 December 1993, as quoted in Larry May, "Challenging Medical Authority: The Refusal of Treatment by Christian Scientists," *Hastings Center Report* 25 (1995): 18.

67. On the use of statistics by Christian Scientists to demonstrate the advantages of practitioner treatment, see Jones, "Child Deaths Put Faith on Trial," 16. For further defenses of Christian Science, see Mark Ruble, "Spiritual Healing for Children," *Update* 8 (1992): 1–2; Allison W. Phinney Jr., "The Spirituality of Mankind," *Christian Science Sentinel*, 3 September 1984, 1529–33; and Peel, *Spiritual Healing*. For examples of suggestions for legislature and courts from the legal community, see Wayne F. Malecha, "Faith Healing Exemptions to Child Protection Laws: Keeping the Faith versus Medical Care for Children," *Journal of Legislation* 12 (1985): 243–63; Jenny Brown, "California Penal Code's Child Neglect / Abandonment Statutes: Religious Freedom or Religious Persecution?" *Santa Clara Law Review* 25 (1985): 613–32; JoAnna A. Gekas, "California's Prayer Healing Dilemma," *Hastings Constitutional Law Quarterly* 14 (1986–87): 395–419; Daniel J. Kearney, "Parental Failure to Provide Child with Medical Assistance Based on Religious Beliefs Causing Child's Death—Involuntary Manslaughter in Pennsylvania," *Dickinson Law Review* 90 (1985–86): 861–90; Donna K. LeClair, "Faith Healing and Religious-Treatment Exemptions to Child-Endangerment Laws: Should Parental Religious Practices Excuse the Failure to Provide Necessary Medical Care to Children?" *University of Dayton Law Review* 13 (1987): 79–106; Ivy B. Dodes, " 'Suffer the Little Children . . .': Toward a Judicial Recognition of a Duty of Reasonable Care Owed Children by Religious Faith Healers," *Hofstra Law Review* 16 (1987): 165–76; Christine A. Clark, "Religious Accommodation and Criminal Liability," *Florida State University Law Review* 17 (1990): 559–90; Paula A. Monopoli, "Allocating the Costs of Parental Free Exercise: Striking a New Balance between Sincere Religious Belief and a Child's Right to Medical Treatment," *Pepperdine Law Review* 18 (1991): 319–52; Edward Egan Smith, "The Criminalization of Belief: When Free Exercise Isn't," *Hastings Law Journal* 42 (1991): 1491–1526; Barry Nobel, "Religious Healing in the Courts: The Liberties and Liabilities of Patients, Parents, and Healers," *University of Puget Sound Law Review* 16 (1993): 599–710; Janet June Anderson, "Capital Punishment of Kids: When Courts Permit Parents to Act on Their Religious Beliefs at the Expense of Their Children's Lives," *Vanderbilt Law Review* 46 (1993): 755–77; Janna C. Mer-

rick, "Christian Science Healing of Minor Children: Spiritual Exemption Statutes, First Amendment Rights, and Fair Notice," *Issues in Law and Medicine* 10 (1994): 321–42; Jennifer L. Rosato, "Putting Square Pegs in a Round Hole: Procedural Due Process and the Effect of Faith Healing Exemptions on the Prosecution of Faith Healing Parents," *University of San Francisco Law Review* 29 (1994): 43–105; Walter Wadlington, "Medical Decision Making for and by Children: Tensions between Parent, State, and Child," *University of Illinois Law Review* (1994): 311–36; and Anne D. Lederman, "Understanding Faith: When Religious Parents Decline Conventional Medical Treatment for Their Children," *Case Western Reserve Law Review* 45 (1995): 891–926.

68. Peel, *Spiritual Healing*, 110; Peel, *Health and Medicine in the Christian Science Tradition*. As I have written elsewhere, *Health and Medicine* reads as "part personal apology for the Christian Science faith, and part jeremiad to a contemporary generation of backslidden Scientists." See *Journal of the American Academy of Religion* 59 (1991): 615.

69. For example, see *Freedom and Responsibility; Christian Science: A Sourcebook of Contemporary Materials* (Boston: Christian Science Publishing Society, 1990); the special issue of the *Christian Science Sentinel* (1991) titled "Humanity's Quest for Health"; and Thomas Johnsen, "Healing and Conscience in Christian Science," *Christian Century* 111 (1994): 640–41.

70. On the chilling effect on Christian Scientists, see the observation of bioethicist Arthur Caplan in Jones, "Child Deaths Put Faith on Trial," 16.

71. Author's enumeration from the authorized listing in the *Christian Science Journal* 118 (2000): 70–86.

72. The biography in question was Bliss Knapp's *The Destiny of the Mother Church* (Boston: Christian Science Publishing Society, 1991), which originally circulated privately under Knapp's 1947 copyright.

73. For details of these crises in finance and church polity, see Peter Steinfels, "Christian Science in Crisis," *Riverside County (Calif.) Press-Enterprise*, 29 February 1992, A-1, A-10; Bob Baker, "Christian Science in Bad Health," *Los Angeles Times*, 2 June 1992, A1, A18; Bob Baker, "Christian Science Financial Ills Disclosed," *Los Angeles Times*, 9 June 1992, A30; "Annual Meeting of the Mother Church—1992," *Christian Science Journal* 110 (1992): 6–40; "Christian Science's Upbeat Financial Data Questioned," *Los Angeles Times*, 28 November 1992, B5; and Stephen Gottschalk, "Christian Science Polity in Crisis," *Christian Century*, 3 March 1993, 242–46, quotation on 244.

Archives Consulted

Archives of the Mother Church, Christian Science Center, Boston
Longyear Museum and Historical Society, Brookline, Massachusetts
Library of Congress, Washington, D.C.
National Library of Medicine, Bethesda, Maryland
Los Angeles County Archives, Hall of Records, Los Angeles, California
Iowa State Historical Department, Division of Historical Museum and Archives,
    Des Moines, Iowa
Kansas State Historical Society, Topeka, Kansas
Maine State Archives, Augusta, Maine
Hennepin County Historical Society, Minneapolis, Minnesota
Missouri Historical Society, St. Louis, Missouri
Missouri Record Management and Archives Service, Kansas City Court of
    Appeals, Transcript and Case Records, Jefferson City, Missouri
Buchanan County Historical Society, St. Joseph, Missouri
St. Joseph Historical Society, St. Joseph, Missouri
Nebraska State Historical Society, Lincoln, Nebraska
Buffalo County Historical Museum, Kearney, Nebraska
New Hampshire Supreme Court Library, Concord, New Hampshire
New Hampshire Department of Administration and Control, Division of
    Records Management and Archives, Concord, New Hampshire
Library of the John Jay College of Criminal Justice, New York, New York
Manuscript Department, Duke University Library, Durham, North Carolina
Supreme Court Law Library, State of Ohio, Columbus, Ohio
Ohio Historical Society, Archives-Manuscripts Division, Columbus, Ohio
Pennsylvania Historical and Museum Commission, Division of Archives and
    Manuscripts, Harrisburg, Pennsylvania
Phillips Memorial Library Archives, Providence College, Providence, Rhode
    Island
Wisconsin State Historical Society, Madison, Wisconsin

Bibliographical Essay

One of the greatest challenges to the study of Christian Science history has been the parsimonious management of the archival resources held in the Christian Science Center in Boston. Given the way critics have often mercilessly vilified the movement in the past, the church's caution can be understood. However, even when the church granted access to responsible scholars to read and take notes, the seemingly endless negotiations (often unsuccessful) for the right to quote fairly from the materials rather than in ways confirming the church's interpretation of the past, or the insistence, in the name of copyright protection, that scholars only paraphrase materials made access seem a pyrrhic victory.

Therefore the Board of Directors' announcement at the church's annual meeting in June 2000 that it intended to house the collections in a newly constructed "Mary Baker Eddy Library for the Betterment of Humanity" and to "permit the orderly, contextual, and publicly accessible presentation" of the collections gave only cautious hope that the modest needs of investigators would be met. And the rationale given for the decision may give one further pause. As reasons, the board cited the public's heightened demand for spiritual answers and increased interest in Eddy; but uppermost may have been the pending change in copyright law that would have made the materials part of the public domain after 2002 if they had not been made publicly accessible by then. And public access through a new library rather than through publication appeared to meet the law's requirements. In this way the church ensures through 2047 that it can continue "to take responsible action to protect and place in context" Eddy's "writings for many years to come" (*Christian Science Journal* 118 [2000]: 54–63).

Effectively denied access to fresh materials, scholars have been challenged to find novel ways to interpret materials previously used in published works or to uncover new public sources. Many of the sources cited below resulted from the former approach, and a few from the latter.

The records of the court trials of Christian Scientists appear to be one of those underutilized sources, and they are fairly easy to track down but moderately difficult to ferret out. The existence of the trials can be determined by references in the *Christian Science Journal* (1883–) and the *Christian Science Sen-*

tinel (1898–), in indexed newspapers and other press, in religious, medical, and legal journals, and in case reporters that document significant state and federal court decisions. Once the cases have been identified and dated, one can locate media and scholarly coverage in nonindexed sources with relative ease.

A well-equipped law library with state and federal reporters and access to online resources such as LEXIS-NEXIS prove invaluable, as do the occasional articles in law journals that give a legal history of religious healing or Christian Science. For excellent overviews and analyses of relevant cases up to the time of their publication dates, see Irving E. Campbell, "Christian Science and the Law," *Virginia Law Register* 10 (1904): 285–300; John C. Myers, "Christian Science and the Law," *Law Notes* 12 (1908–9): 5–6; Peter V. Ross, "Metaphysical Treatment of Disease as the Practice of Medicine," *Yale Law Journal* 24 (1914–15): 391–411; and I. H. Rubenstein, "Faith Healing, Legal Aspects," in *A Treatise on Contemporary Religious Jurisprudence* (Chicago: Waldain Press, 1948). For outstanding introductions to the legal sources and issues involving children and religious healing, see Elena M. Kondos, "The Law and Christian Science Healing for Children: A Pathfinder," *Legal Reference Services Quarterly* 12 (1992): 5–71, and Rita Swan, "Children, Medicine, Religion, and the Law," *Advances in Pediatrics* 44 (1997): 491–543.

Determining whether unpublished materials related to the trials still exist, and locating them if they do, can be more difficult. Although federal law mandates the preservation and maintenance of legal records, Congress leaves it to the states to allocate funds for their own judicial systems. Therefore in some states one finds court records—from indictments to briefs and trial transcripts—safely reposing and well accessioned, while in other jurisdictions a city or county clerk takes one to a pile of boxes in the basement. References to the materials I have located appear in the notes of the appendix.

Several kinds of sources prove useful in the analysis and interpretation of the cases. Articles in legal journals not only discuss the particulars of various cases but reflect on the legal principles involved and offer judicial or legislative remedies for the issues raised by the cases. For a detailed analysis of the rhetorical patterns of famous trials, see Janice Schuetz and Kathryn Holmes Snedaker, *Communication and Litigation: Case Studies of Famous Trials* (Carbondale: Southern Illinois University Press, 1988), and W. Lance Bennett and Martha S. Feldman, *Reconstructing Reality in the Courtroom: Justice and Judgment in American Culture* (New Brunswick, N.J.: Rutgers University Press, 1981). For discussion of the sociopolitical nature of trials widely covered by the media—so-called popular trials—see Michael R. Belknap, ed., *American Political Trials* (Westport, Conn.: Greenwood Press, 1981). For the role of the press in retelling the story of trials to the public, see Robert Hariman, ed., *Popular Trials: Rhetoric, Mass Media, and the Law* (Tuscaloosa: University of Alabama Press, 1990), especially Lawrence M. Bernabo and Celeste Michelle Condit, "Two Stories of the Scopes Trial: Legal and Journalistic Articulations of the Legitimacy of Science and Religion," 55–85, 204–18, and John Lofton, *Justice and the Press* (Boston: Beacon Press, 1966).

Regina Markell Morantz-Sanchez has demonstrated the effectiveness of several of these critical approaches in *Conduct Unbecoming a Woman: Medicine on Trial in Turn-of-the-Century Brooklyn* (New York: Oxford University Press, 1999).

Since Christian Science exhibited characteristics of both a medical and a religious movement, analyses of its beliefs and practices usually have come with a bias for one or the other. For a focus on the religious teachings of the movement, see Stephen Gottschalk's *The Emergence of Christian Science in American Religious Life* (Berkeley: University of California Press, 1973). For an emphasis on medical healing, see Rennie B. Schoepflin, "Christian Science Healing in America," in *Other Healers: Unorthodox Medicine in America*, ed. Norman Gevitz (Baltimore: Johns Hopkins University Press, 1988), 192–214.

Finally, an understanding of the intersection of medicine and religion in America provides a necessary grounding for the study of Christian Science; numerous sources document these relationships. For example, consult *Caring and Curing: Health and Medicine in the Western Religious Traditions*, ed. Ronald L. Numbers and Darrel W. Amundsen (New York: Macmillan, 1986, reprint, Johns Hopkins University Press, 1997); "Part IX. Medicine and Psychology," in *The History of Science and Religion in the Western Tradition: An Encyclopedia*, ed. Gary B. Ferngren, Edward J. Larson, and Darrel W. Amundsen (New York: Garland, 2000); and "America's Innovative Nineteenth-Century Religions," in *History of Science and Religion*, 307–12. On antebellum medical ties between science and religion, see Steven M. Stowe, "Religion, Science, and the Culture of Medical Practice in the American South, 1800–1870," in *The Mythmaking Frame of Mind: Social Imagination and American Culture*, ed. James Gilbert et al. (Belmont, Calif.: Wadsworth, 1993), 1–24; and Charles E. Rosenberg, *The Cholera Years: The United States in 1832, 1849, and 1866* (Chicago: University of Chicago Press, 1962). For insightful views of the long-standing debate over miracle, prayer, and healing, see Robert Bruce Mullin, *Miracles and the Modern Religious Imagination* (New Haven: Yale University Press, 1996), and Rick Ostrander, *The Life of Prayer in a World of Science: Protestants, Prayer, and American Culture, 1870–1930* (New York: Oxford University Press, 2000). For a recently resurgent interest in medicine and prayer, see Larry Dossey, *Healing Words: The Power of Prayer and the Practice of Medicine* (San Francisco: HarperCollins, 1993), and the wide-ranging responses it has received.

The published scholarship on Mary Baker Eddy and the Christian Science movement is extensive; what follows includes works that prove especially useful in developing an understanding of the intersection of religion and medicine in her life. Many authors interested in Christian Science have allowed Mary Baker Eddy to overpower their work, and critical, often unflattering accounts of Eddy and her movement have set the agenda for much of the work done by historians and social scientists. Mark Twain's *Christian Science* (New York: Harper, 1907), Georgine Milmine's serial exposé in *McLure's Magazine* (vol. 28, no. 2 [December 1906] to vol. 31, no. 2 [June 1908]), published as *The Life of Mary G. Baker Eddy and the History of Christian Science* (New York: Doubleday, Page, 1909), recently reissued as Willa Cather and Georgine Milmine, *The Life of Mary Baker G. Eddy and the History of Christian Science* (Lincoln: University of Nebraska Press, 1993), and Edwin Franden Dakin's *Mrs. Eddy: The Biography of a Virginal Mind* (New York: Charles Scribner's Sons, 1929) have shown the way. Beyond their often unflattering accounts of the founder, however, these early critics document the often contentious nature of her flock and wryly note the material well-being of many of her idealist followers. Ernest Sutherland Bates and John V. Dittemore expand these themes in *Mary Baker Eddy: The Truth and the Tradition* (New York: Alfred A. Knopf, 1932), a particularly good source for primary materials and organizational details of early Christian Science.

As one might expect, Christian Science nurtured a hagiographical and triumphalist tradition that drew unquestioningly from Eddy's own writings and reminiscences and unpublished manuscripts, including her autobiography, *Retrospection and Introspection* (Boston, 1891). Prototypical of such efforts is Sibyl Wilbur's inspirational and ethereal *Life of Mary Baker Eddy* (New York: Concord, 1908), which stood for years as the church's semiofficial image; Lyman P. Powell, *Mary Baker Eddy: A Life Size Portrait* (New York: Macmillan, 1930), and Norman Beasley, *Mary Baker Eddy* (New York: Duell, Sloan, and Pearce, 1963).

Often as an alternative to Christian Scientists' simple trust in Eddy's divine inspiration and the hyperbole of many critics, scholars turned to psychological and medical models to understand her personality. First Dakin, then Julius Silberger Jr., *Mary Baker Eddy: An Interpretive Biography of the Founder of Christian Science* (Boston: Little, Brown, 1980), and finally Robert David Thomas, *"With Bleeding Footsteps": Mary Baker Eddy's Path to Religious Leadership* (New York: Alfred A. Knopf, 1994), identify similar psychoanalytic themes and conclude in their psychobiographies that Eddy possessed a complex, if troubled and often contradictory, self that she used to build her own successes. Noteworthy among the numerous other sources that examine the mind of Eddy are Stefan Zweig, *Mental Healers: Anton Mesmer, Mary Baker Eddy, Sigmund Freud* (Garden City, N.Y.: Garden City, 1932); Gail Thain Parker, "Mary Baker Eddy: New Thought Parodied," in *Mind Cure in New England: From the Civil War to World War I* (Hanover, N.H.: University Press of New England, 1973), 109–29; George Pickering, *Creative Malady: Illness in the Lives and Minds of Charles Darwin, Florence Nightingale, Mary Baker Eddy, Sigmund Freud, Marcel Proust, Elizabeth Barrett Browning* (New York: Oxford University Press, 1974); F. E. Kenyon, "Mary Baker Eddy, Founder of Christian Science: The Sublime Hysteric," *History of Medicine* (London) 6 (1975): 29–46; John K. Maniha and Barbara B. Maniha, "A Comparison of Psychohistorical Differences among Some Female Religious and Secular Leaders," *Journal of Psychohistory* 5 (1978): 523–49; and Janice Klein, "Ann Lee and Mary Baker Eddy: The Parenting of New Religions," *Journal of Psychohistory* 6 (1979): 361–75.

We still lack a sophisticated analysis of the evolution of Eddy's thought that systematically examines the various key editions (most notably first, third, fourteenth, twenty-second, and fiftieth) of *Science and Health*. For simple documentation of change see [Alice L. Orgain], *Distinguishing Characteristics of Mary Baker Eddy's Progressive Revisions of "Science and Health" and Other Writings* (New York: Rare Book Company, 1933), and William Dana Orcutt, *Mary Baker Eddy and Her Books* (Boston: Christian Science Publishing Society, 1950). The study of the real and imagined literary relationships between Eddy, Phineas Parkhurst Quimby, and others forms its own cottage industry. Horatio W. Dresser's defense of Quimby's originality appears as *The Quimby Manuscripts Showing the Discovery of Spiritual Healing and the Origin of Christian Science* (New York: Thomas Y. Crowell, 1921), and for a defense of Eddy against charges that she stole Christian Science from Quimby, see Thomas C. Johnsen, "Historical Consensus and Christian Science: The Career of a Manuscript Controversy," *New England Quarterly* 53 (March 1980): 3–22. But the best single source for examining such issues of supposed literary dependency on Quimby is Ervin Seale's three-volume edition of Quimby's manuscripts, *Phineas Parkhurst Quimby: The Complete Writings* (Marina del Rey, Calif.: Devorss, 1988). For an insightful analysis of *Science and Health* as the locus of authority in Christian Science, see David L. Weddle, "The Christian Science Textbook: An Analysis of the Religious Authority of *Science and Health* by Mary Baker Eddy," *Harvard Theological Review* 84 (1991): 273–97.

Works that recreate the world of Eddy and explore her relationships with her contemporary movements form another body of scholarship. In Charles Braden's seminal study of American religious cults, *These Also Believe: A Study of Modern American Cults and Minority Religious Movements* (New York: Macmillan, 1949), the chapter titled "Christian Science" (180–220) examines the intellectual roots of Eddy's teachings, the institutionalization of her movement, and the beliefs and practices of Christian Scientists. Robert Peel places Christian Science in the context of transcendentalism in *Christian Science: Its Encounter with American Culture* (New York: Henry Holt, 1958) and concludes that they represented "related but distinct" intellectual traditions; and in his pathbreaking three-volume biography *Mary Baker Eddy* (New York: Holt, Rinehart, Winston, 1966–77) he delivers a comprehensive overview of Eddy and her world in which he builds on the social theory of sects and divides Eddy's life into three crucial phases: "prophetic" discovery, struggle for legitimacy, and routinization of authority. Although Peel often puts the best face on the many troublesome episodes of Eddy's career, he obliterates the myth of a changeless Eddy and replaces it with a picture of a creative and adaptive woman influenced by homeopathy, Quimbyism, and the women's movement and firmly grounded in the intellectual and cultural traditions of New England.

Feminist analysis of Eddy appears in Jean A. McDonald, "Mary Baker Eddy and the Nineteenth-Century 'Public' Woman: A Feminist Reappraisal," *Journal of Feminist Studies* 2 (1986): 105–11; and in an insightful feminist defense, *Mary Baker Eddy* (Reading, Mass.: Perseus Books, 1998), biographer Gillian Gill concludes that much of the criticism Eddy and her movement endured arose from the fact that she refused to conform to the expected roles for public women in her time.

Nineteenth-century America spawned many female prophets and ecstatics, some of whom connected religion with mental or physical health. Ronald L. Numbers's religiomedical biography of Ellen G. White, founder of Seventh-day Adventism, early set the standard for such studies: *Prophetess of Health: A Study of Ellen G. White* (New York: Harper and Row, 1976; see also the revised and enlarged edition (Knoxville: University of Tennessee Press, 1992). Ronald L. Numbers and Rennie B. Schoepflin compare and contrast the efforts of White and Eddy to "found their social niche[s] at the intersection of medical and religious reform" in "Ministries of Healing: Mary Baker Eddy, Ellen G. White, and the Religion of Health," in *Women and Health in America: Historical Readings*, 2d ed., ed. Judith Walzer Leavitt (Madison: University of Wisconsin Press, 1999), 579–95. See also Joel Whitney Tibbetts, "Women Who Were Called: A Study of the Contributions to American Christianity of Ann Lee, Jemima Wilkinson, Mary Baker Eddy and Aimee Semple McPherson" (Ph.D. diss., Vanderbilt University, 1976), and the bibliographical essay by Jonathan M. Butler and Rennie B. Schoepflin, "Charismatic Women and Health: Mary Baker Eddy, Ellen G. White, and Aimee Semple McPherson," in *Women, Health, and Medicine: A Historical Handbook*, ed. Rima D. Apple (New York: Garland, 1990): 337–65 [reissued New Brunswick, N.J.: Rutgers University Press, 1992]: 329–57).

Eddy explored several alternative healing pathways, whose ideologies and adherents often intertwined in the nineteenth century, before she settled on Quimbyism and then Christian Science: spiritualism, phrenology, hydropathy, homeopathy, and her bête noire, mesmerism. For striking parallels between aspects of early Christian Science healing and healing mediums, see R. Laurence Moore, *In Search of White Crows: Spiritualism, Parapsychology, and American Culture* (New York: Oxford University Press, 1977); R. Laurence Moore, "The Occult Connection? Mormonism, Christian Science, and Spiritualism," in *The Occult in America: New Historical Perspectives*, ed. Howard Kerr and Charles L. Crow (Urbana: University of Illinois Press, 1983), 135–61; Ann Braude, *Radical Spirits: Spiritualism and Women's Rights in Nineteenth-Century America* (Boston: Beacon Press, 1989), 182–89; and Ann Taves, *Fits, Trances, and Visions: Experiencing Religion and Explaining Experience from Wesley to James* (Princeton: Princeton University Press, 1999), 215–19. For overviews of phrenology consult John D. Davies, *Phrenology—Fad and Science: A Nineteenth-Century American Crusade* (New Haven: Yale University Press, 1955), and Madeleine B. Stern, *Heads and Headlines: The Phrenological Fowlers* (Norman: University of Oklahoma Press, 1971). On hydropathy and its female connections, see Susan E. Cayleff, "Gender, Ideology, and the Water-Cure," in *Other Healers: Unorthodox Medicine in America*, ed. Norman Gevitz (Baltimore: Johns Hopkins University Press, 1988), 82–98; Susan E. Cayleff, *Wash and Be Healed: The Water-Cure Movement and Women's Health* (Philadelphia: Temple University Press, 1987); and Jane B. Donegan, *"Hydropathic Highway to Health": Women and Water-Cure in Antebellum America* (Westport, Conn.: Greenwood Press, 1986). On homeopathy, see Martin Kaufman, *Homeopathy in America: The Rise and Fall of a Medical Heresy* (Baltimore:

Johns Hopkins University Press, 1971), and Anne Taylor Kirschmann, "Adding Women to the Ranks, 1860–1890: A New View with a Homeopathic Lens," *Bulletin of the History of Medicine* 73 (1999): 429–46. For mesmerism and its relation to healing, see Robert C. Fuller, *Mesmerism and the American Cure of Souls* (Philadelphia: University of Pennsylvania Press, 1982); Peter McCandless, "Mesmerism and Phrenology in Antebellum Charleston: 'Enough of the Marvellous,' " *Journal of Southern History* 58 (1992): 199–230; and Mary Farrell Bednarowski, "Women in Occult America," in *The Occult in America: New Historical Perspectives*, ed. Howard Kerr and Charles L. Crow (Urbana: University of Illinois Press, 1983): 177–95.

Finally, for an insightful overview of the variety of similar ways nineteenth-century Americans blended medical and religious sectarianism and often moved toward metaphysical healing, see Donald Meyer, *The Positive Thinkers: A Study of the American Quest for Health, Wealth and Personal Power from Mary Baker Eddy to Norman Vincent Peale* (New York: Random House, 1965; reissued under a slightly altered title in 1980); Gail Thain Parker, *Mind Cure in New England: From the Civil War to World War I* (Hanover, N.H.: University Press of New England, 1973); Mary Farrell Bednarowski, "Outside the Mainstream: Women's Religion and Women Religious Leaders in Nineteenth-Century America," *Journal of the American Academy of Religion* 48 (1980): 207–31; Robert C. Fuller, *Alternative Medicine and American Religious Life* (New York: Oxford University Press, 1989); and Catherine L. Albanese, *Nature Religion in America: From the Algonkian Indians to the New Age* (Chicago: University of Chicago Press, 1990).

Christian Science attracted both men and women, but in many ways it was a woman's movement, founded by and committed to women. Sources that develop that female connection include Margery Q Fox, "Power and Piety: Women in Christian Science" (Ph.D. diss., New York University, 1973); Gage William Chapel, "Christian Science and the Nineteenth Century Woman's Movement," *Central States Speech Journal* 26 (1975): 142–49; and Penny Hansen, "Woman's Hour: Feminist Implications of Mary Baker Eddy's Christian Science Movement, 1885–1910" (Ph.D. diss., University of California, Irvine, 1981).

Many Christian Scientists reported that metaphysical healing effectively eased the difficulties of childbirth. For a clearer understanding of the pain and danger that often accompanied childbirth, see Judith Walzer Leavitt, *Brought to Bed: Childbearing in America, 1750–1950* (New York: Oxford University Press, 1986), and Richard W. Wertz and Dorothy C. Wertz, *Lying-In: A History of Childbirth in America* (New York: Schocken Books, 1977). For glimpses into the life of a midwife or obstetrician, see Laurel Thatcher Ulrich, *A Midwife's Tale: The Life of Martha Ballard, Based on Her Diary, 1785–1812* (New York: Vintage Books, 1990), 162–203; Charlotte G. Borst, *Catching Babies: The Professionalization of Childbirth, 1870–1920* (Cambridge: Harvard University Press, 1996); and Steven M. Stowe, "Obstetrics and the Work of Doctoring in the Mid-Nineteenth-Century American South," *Bulletin of the History of Medicine* 64 (1990): 540–66.

In many ways the metaphysical worldview of Christian Scientists rendered them foreigners in their own land. For helpful cross-cultural views of healing and an outstanding bibliography, see Arthur Kleinman, *Patients and Healers in the Context of Culture: An Exploration of the Borderland between Anthropology, Medicine, and Psychiatry* (Berkeley: University of California Press, 1980). For a recent example of worldviews clashing over medicine, religion and epilepsy, see Anne Fadiman, *The Spirit Catches You and You Fall Down: A Hmong Child, Her American Doctors, and the Collision of Two Cultures* (New York: Farrar, Straus, and Giroux, 1997). Regarding women healers, see Carol Shepherd McClain, ed., *Women as Healers: Cross-Cultural Perspectives* (New Brunswick, N.J.: Rutgers University Press, 1989), which includes Margery Fox's article "The Socioreligious Role of the Christian Science Practitioner," 98–114. Cynthia Grant Tucker enlivens our appreciation of the experiences of a Christian Science practitioner who lived from 1870 to 1953 in *A Woman's Ministry: Mary Collson's Search for Reform as a Unitarian Minister, a Hull House Social Worker, and a Christian Science Practitioner* (Philadelphia: Temple University Press, 1984), reissued as *Healer in Harm's Way: Mary Collson, A Clergywoman in Christian Science* (Knoxville: University of Tennessee Press, 1994). Essential biographical facts about the lives of early believers included in *Pioneers in Christian Science* (Brookline, Mass.: Longyear Historical Society, 1972, rev. 1980) are developed in greater detail in the numerous articles that have appeared in the Longyear Historical Society and Museum's *Quarterly News* (1964–).

Christian Science healers operated in the context of many social and economic circumstances. For introductions to women and health in America, see *Women and Health in America: Historical Readings*, 2d ed., ed. Judith Walzer Leavitt (Madison: University of Wisconsin Press, 1999), and *Women, Health, and Medicine* (cited above). For reflections on gender, class, and medical pluralism in America, see Hans A. Baer, "The American Dominative Medical System as a Reflection of Social Relations in the Larger Society," *Social Science and Medicine* 28 (1989): 1103–12, and Hans A. Baer, *Biomedicine and Alternative Healing Systems in America: Issues of Class, Race, Ethnicity, and Gender* (Madison: University of Wisconsin Press, 2001). For a discussion of the female attraction to sectarian medicine and its domestic nature, see Naomi Rogers, "Women and Sectarian Medicine," in *Women, Health, and Medicine* (cited above; 1990: 281–310, 1992: 273–302). For comparative purposes, consult a firsthand account and analysis of nonsectarian, female family members "doctoring" their households in Emily K. Abel, "Family Caregiving in the Nineteenth Century: Emily Hawley Gillespie and Sarah Gillespie, 1858–1888," *Bulletin of the History of Medicine* 68 (1994): 573–99. On the difficulty many women had in gaining access to formal medical education, see Mary Roth Walsh, *"Doctors Wanted, No Women Need Apply": Sexual Barriers in the Medical Profession, 1835–1975* (New Haven: Yale University Press, 1977). For an extended discussion of the ways work and community transformed the lives of single women, see Martha Vicinus, *Independent Women: Work and Community for Single Women* (Chicago: University of Chicago Press, 1985).

The desire to be healed, to heal, and to teach about healing each played a role in stimulating conversion to Christian Science. Despite their apologetic nature, the testimonials published in the *Christian Science Journal* reveal much about what attracted an individual to Christian Science. See, for example, the analysis of testimonies by R. W. England, "Some Aspects of Christian Science as Reflected in Letters of Testimony," *American Journal of Sociology* 59 (March 1954): 448–53, and Penny Hansen, "Woman's Hour" (cited above). For an analysis of the attraction of Christian Science as a popular religion that offered "solution to the problem of the emotional dissatisfactions of everyday life," see R. Laurence Moore, "Christian Science and American Popular Religion," in *Religious Outsiders and the Making of Americans* (New York: Oxford University Press, 1986), 105–27.

Several studies clarify the demographic characteristics of Christian Science. Neal B. DeNood discusses the spread of Christian Science beyond its New England origins in "The Diffusion of a System of Belief" (Ph.D. diss., Harvard University, 1937); Harold W. Pfautz analyzes the urban characteristics of Christian Science in "A Case Study of an Urban Religious Movement: Christian Science," in *Contributions to Urban Sociology*, ed. Ernest W. Burgess and Donald J. Bogue (Chicago: University of Chicago Press, 1964), 284–303; and the best overall source is Ary Johannes Lamme III, "The Spatial and Ecological Characteristics of the Diffusion of Christian Science in the United States: 1875–1910" (D.S.S. diss., Syracuse University, 1968).

Religious healing has manifested itself in virtually all times and cultures, often functioning simultaneously as religion and medicine. For two outstanding twentieth-century examples of healing rituals as an expression of personal piety, see two books by Robert A. Orsi, *Thank You, St. Jude: Women's Devotion to the Patron Saint of Hopeless Causes* (New Haven: Yale University Press, 1996), and *The Madonna of 115th Street: Faith and Community in Italian Harlem, 1880–1950* (New Haven: Yale University Press, 1985). The best introduction to healing in the American Pentecostal traditions is David Edwin Harrell Jr., *All Things Are Possible: The Healing and Charismatic Revivals in Modern America* (Bloomington: Indiana University Press, 1975); see also Edith L. Blumhofer, *Aimee Semple McPherson: Everybody's Sister* (Grand Rapids: William B. Eerdmans, 1993).

Overviews of the beliefs and practices of Christian Science healers appear in Walter I. Wardwell, "Christian Science Healing," *Journal for the Scientific Study of Religion* 4 (spring 1965): 175–81; Arthur Corey, *Behind the Scenes with the Metaphysicians* (Los Angeles: DeVorss, 1968); and Rennie B. Schoepflin, "The Christian Science Tradition," in *Caring and Curing: Health and Medicine in the Western Religious Traditions*, ed. Ronald L. Numbers and Darrel W. Amundsen (New York: Macmillan, 1986; reprint, Johns Hopkins University Press, 1997), 421–46. The theology of Christian Science healing is explored comparatively in Mary Farrell Bednarowski, *New Religions and the Theological Imagination in America* (Bloomington: Indiana University Press, 1989), and Gottschalk, *Emergence of Christian Science*, 222–38. Lucy Jayne Kamau explores its ritual dimensions in "Systems of Belief and Ritual in Christian Science" (Ph.D. diss., University of Chicago, 1971). For official church views of the centrality of healing to a Christian

Science life, see *A Century of Christian Science Healing* (Boston: Christian Science Publishing Society, 1966). For a better understanding of the fact that all disease etiologies (not just the metaphysical ones of Christian Science) are culture bound, see Charles E. Rosenberg and Janet Golden, eds., *Framing Disease: Studies in Cultural History* (New Brunswick, N.J.: Rutgers University Press, 1992).

Under the influence of a new generation of social historians and sociologists in the 1950s, scholars who studied Christian Science shifted their focus from Eddy not only to the beliefs and practices of Christian Scientists, but also to an analysis of the structure of their organizations. For early "insider" examples of this trend, see Norman Beasley, *The Cross and the Crown: The History of Christian Science* (New York: Duell, Sloan and Pearce, 1952), and Norman Beasley, *The Continuing Spirit: The Story of Christian Science since 1910* (New York: Duell, Sloan and Pearce, 1956). "Outsider" Charles Braden looks for similar cultural parallels and finds them in the plethora of late nineteenth-century mind healing movements that collectively became New Thought (*Spirits in Rebellion: The Rise and Development of New Thought* [Dallas: Southern Methodist University Press, 1963]). In *Positive Thinkers* Donald Meyer (see above) concludes that Christian Science attracted many followers because of the power over self that it and similar pop psychologies from Dale Carnegie to Oral Roberts offered to Americans. Specific details and individuals involved in these movements can be explored through a study of the several journals published by the mental healers, mind healers, occult healers, and "generic" Christian Scientists of the Progressive Era (see chapter 4 above). See also Gail M. Harley's detailed study "Emma Curtis Hopkins: 'Forgotten Founder' of New Thought" (Ph.D. diss., Florida State University, 1991).

As revealed in their medical journals, Christian Scientists were not the only unorthodox healing group to have rocky relationships with M.D.s. Physicians attacked osteopaths and chiropractors during the same period. See Norman Gevitz, *The D.O.'s: Osteopathic Medicine in America* (Baltimore: Johns Hopkins University Press, 1982), and J. Stuart Moore, *Chiropractic in America: The History of a Medical Alternative* (Baltimore: Johns Hopkins University Press, 1993). For views of the relationship between Christian Science practitioners and the medical profession, see Margery Fox, "Conflict to Coexistence: Christian Science and Medicine," *Medical Anthropology* 8 (1984): 292–301, and Thomas C. Johnsen, "Christian Scientists and the Medical Profession: A Historical Perspective," *Medical Heritage* 2 (January–February 1986): 70–78.

For helpful views of the psychological developments of the late nineteenth and early twentieth centuries and their relation to religion, see John Chynoweth Burnham, *Psychoanalysis and American Medicine, 1894–1918: Medicine, Science, and Culture*, Psychological Issues, Monograph 20 (New York: International Universities Press, 1967); Nathan G. Hale Jr., *Freud and the Americans: The Beginnings of Psychoanalysis in the United States, 1876–1917* (New York: Oxford University Press, 1971), and his *The Rise and Crisis of Psychoanalysis in the United States: Freud and the Americans, 1917–1985* (New York: Oxford University Press, 1995); Eric

Caplan, *Mind Games: American Culture and the Birth of Psychotherapy* (Berkeley: University of California Press, 1998); and Eugene Taylor, *Shadow Culture: Psychology and Spirituality in America* (Washington, D.C.: Counterpoint, 1999).

The Emmanuel movement sought to merge the best of scientific medicine with the best of religious healing. For detailed discussions of that movement, see Raymond Joseph Cunningham, "The Emmanuel Movement: A Variety of American Religious Experience," *American Quarterly* 14 (1962): 48–63; Raymond Joseph Cunningham, "Ministry of Healing: The Origins of the Psychotherapeutic Role of the American Churches" (Ph.D. diss., Johns Hopkins University, 1965), esp. chaps 4–5, pp. 113–89; Raymond J. Cunningham, "From Holiness to Healing: The Faith Cure in America," *Church History* 43 (1974): 499–513; and Sanford Gifford, "Medical Psychotherapy and the Emmanuel Movement in Boston, 1904–1912," in *Psychoanalysis, Psychotherapy and the New England Medical Scene, 1894–1944*, ed. George E. Gifford Jr. (New York: Science History Publications/USA, 1978), 106–18.

For discussion of ways the established American denominations experienced a renaissance of faith healing in response to Christian Science, see Raymond J. Cunningham, "The Impact of Christian Science on the American Churches, 1880–1910," *American Historical Review* 72 (1967): 885–905, and Robert Bruce Mullin, "The Debate over Religion and Healing in the Episcopal Church: 1870–1930," *Anglican and Episcopal History* 60 (1991): 213–34.

American medicine underwent monumental ideological, therapeutic, and professional change during the Progressive Era, and the uncertainties that accompanied those changes exacerbated the tensions with Christian Scientists. John Harley Warner documents many of the therapeutic changes that transformed medicine on the eve of the new century in *The Therapeutic Perspective: Medical Practice, Knowledge, and Identity in America, 1820–1885* (Cambridge: Harvard University Press, 1986). On the national debates over science and its authority in public life, which formed one of the backdrops for those discussions, see John C. Burnham on the popularization of science and health in America in *How Superstition Won and Science Lost: Popularizing Science and Health in the United States* (New Brunswick, N.J.: Rutgers University Press, 1987), esp. chap. 2, pp. 45–84.

When physicians moved to legally establish their professional authority, they often worked legally and legislatively. James C. Mohr explores legal issues in *Doctors and the Law: Medical Jurisprudence in Nineteenth-Century America* (New York: Oxford University Press, 1993), and Harold Walter Eickhoff details the legislative efforts in one state in "The Organization and Regulation of Medicine in Missouri, 1883–1901" (Ph.D. diss., University of Missouri, 1964).

The effort to establish and control medical licensing constituted one of the major battlefields in the struggle for medical authority. For an overview of medical licensing in America during this period, see Samuel L. Baker, "Physician Licensure Laws in the United States, 1865–1915," *Journal of the History of Medicine* 39 (1984): 173–97; Samuel L. Baker, "A Strange Case: The Physician Licensure Campaign in Massachusetts in 1880," *Journal of the History of Medicine* 40 (1985): 286–308; and William G. Rothstein, *American Physicians in the Nineteenth Century: From Sects to Science* (Baltimore: Johns Hopkins University Press, 1967), 305–10. For the influence of educational reforms on licensing, see Richard Harrison Shryock, *Medical Licensing in America, 1650–1965* (Baltimore: Johns Hopkins University Press, 1967). For the influence of scientific medicine and the merger of the sects on licensing, see Rothstein, *American Physicians*. To view medical licensure in the general context of occupational licensing, see Lawrence M. Friedman, "Freedom of Contract and Occupational Licensing, 1890–1910: A Legal and Social Study," *California Law Review* 53 (1965): 487–534. For a helpful view of the role that both occupational power and state governments played in the expansion of occupational licensing, see Xueguang Zhou, "Occupational Power, State Capacities, and the Diffusion of Licensing in the American States: 1890 to 1950," *American Sociological Review* 58 (1993): 536–52. See also Samuel Haber, *The Quest for Authority and Honor in the American Professions, 1750–1900* (Chicago: University of Chicago Press, 1991), 319–58.

For the view that physicians sought a competitive advantage through licensing, see Paul Starr, *The Social Transformation of American Medicine* (New York: Basic Books, 1982), 102–12. For the view that they sought a monopoly, see Rothstein, *American Physicians*, and E. Richard Brown, *Rockefeller Medicine Men: Medicine and Capitalism in America* (Berkeley: University of California Press, 1979).

For a good introduction to the issues involved when marginal religious and medical communities become entangled with the state, see the sociopolitical analysis of Fred M. Frohock, *Healing Powers: Alternative Medicine, Spiritual Communities, and the State* (Chicago: University of Chicago Press, 1992). On the ways Christian Scientists negotiated relationships with the state, see Lee Zeunert Johnson, "Christian Science Committee on Publication: A Study of Group and Press Interaction" (Ph.D. diss., Syracuse University, 1963).

One of the key contexts for understanding the conflicts between Scientists and physicians is urban public health. Judith Walzer Leavitt explores the influence of the politics of race and class on public health in *The Healthiest City: Milwaukee and the Politics of Health Reform* (Princeton: Princeton University Press, 1982). On the relation between filth, garbage, and disease, see Martin V. Melosi, *Garbage in the Cities: Refuse, Reform, and the Environment, 1880–1980* (College Station: Texas A&M University Press, 1981). Regarding personal hygiene and public health, see Marilyn Thornton Williams, *Washing "the Great Unwashed": Public Baths in Urban America, 1840–1920* (Columbus: Ohio State University Press, 1991). On the interplay among hygiene, germs, and public health, see Nancy Tomes, *The Gospel of Germs: Men, Women, and the Microbe in American Life* (Cambridge: Harvard University Press, 1998). See also Barbara Gutmann Rosenkrantz, *Public Health and the State: Changing Views in Massachusetts, 1842–1936* (Cambridge: Harvard University Press, 1972), and her "Cart before Horse: Theory, Practice and Professional Image in American Public Health, 1870–1920," *Journal of the History of Medicine* 29 (1974): 66–70. A recurring tension between such public health efforts and Christian Science healing arose as a result of the clash between individual freedom and public legislation. For an early twentieth-century story about the conflict between individual liberty and public health that did not involve Scientists, see Judith Walzer Leavitt, *Typhoid Mary: Captive to the Public's Health* (Boston: Beacon Press, 1996).

Diphtheria played a decisive role in the efforts to curtail and control religious healers, and disputes often arose regarding the best way to confront the disease. On the nature and treatment of diphtheria, see Anne Hardy, "Tracheotomy versus Intubation: Surgical Intervention in Diphtheria in Europe and the United States, 1825–1930," *Bulletin of the History of Medicine* 66 (1992): 536–59, and Charles R. King, *Children's Health in America: A History* (New York: Twayne, 1993): 76–79. Regarding the varied rates of decline in diphtheria death rates, see Gretchen A. Condran, Henry Williams, and Rose A. Cheney, "The Decline in Mortality in Philadelphia from 1870 to 1930: The Role of Municipal Services," in *Sickness and Health in America: Readings in the History of Medicine and Public Health*, 3d ed. rev., ed. Judith Walzer Leavitt and Ronald L. Numbers (Madison: University of Wisconsin Press, 1997), 452–66. On the growing support by physicians of antitoxin, see Rosenkrantz, "Cart before Horse," 69–72 (above).

The changing social and legal standing and medical status of children at the turn of the century were central to debates over healing. See Mary Ann Mason, *From Father's Property to Children's Rights: The History of Child Custody in the United States* (New York: Columbia University Press, 1994); Joseph M. Hawes, *The Children's Rights Movement: A History of Advocacy and Protection* (Boston: Twayne, 1991), esp. chaps. 2–3; and Walter I. Trattner, *From Poor Law to Welfare State: A History of Social Welfare in America*, 2d ed. (New York: Free Press, 1979), esp. chaps. 5–6. On the origins of pediatrics as an American medical specialty, see Sydney A. Halpern, *American Pediatrics: The Social Dynamics of Professionalism, 1880–1980* (Berkeley: University of California Press, 1988), esp. chaps. 3–5.

Reflections on the evolving nature of Christian Science after the death of Eddy are part of most biographies of her, and they appear in most institutional histories that venture beyond 1910. However, of particular note are Bryan Wilson's *Sects and Society: A Sociological Study of the Elim Tabernacle, Christian Science, and Christadelphians* (Berkeley: University of California Press, 1961), esp. chaps. 6–10, pp. 121–215, and Charles S. Braden's *Christian Science Today: Power, Policy, Practice* (Dallas: Southern Methodist University Press, 1958), which deal with the routinization of Eddy's authority through the organization of the Christian Science church and institutions such as the Christian Science Publishing Society and the Board of Directors of the Mother Church. For further insight into the process of adapting religious movements to a dead prophet, see Timothy Miller, *When Prophets Die: The Postcharismatic Fate of New Religious Movements* (Albany: State University of New York Press, 1991). See also Harold W. Pfautz, who used the church as a case study of the secularization of a religious movement in "Christian Science: The Sociology of a Social Movement and a Religious Group" (Ph.D. diss., University of Chicago, 1954) and in "Christian Science: A Case Study of the Social Psychological Aspect of Secularization," *Social Forces* 34 (March 1956): 246–51.

The health practices of contemporary believers are documented in Arthur Edmund Nudelman, "Christian Science and Secular Medicine" (Ph.D. diss., University of Wisconsin, 1970). Robert Peel's *Health and Medicine in the Christian Science Tradition: Principle, Practice, and Challenge* (New York: Crossroad, 1988), part apology for Christian Science and part jeremiad to contemporary Scientists, offers a distinctive look at the movement from the perspective of the troubled 1980s.

Two noteworthy published sources detail the events surrounding late twentieth-century trials of Christian Scientists and other religious healers. Leo Damore tells the troubling story of Dorothy Sheridan and its aftermath in Leo Damore's *The "CRIME" of Dorothy Sheridan* (New York: Arbor House, 1978), and Richard J. Brenneman details the travails of Seth Glaser in *Deadly Blessings: Faith Healing on Trial* (Buffalo, N.Y.: Prometheus Books, 1990), chapter titled "Nestling's Faltering Flight: The Short Life and Death of Seth Ian Glaser."

Index

Eddy's authority, 106; healing role of, 36–37, 87, 149; healing testimonies in, 33–34, 35–36, 38, 75; and medical boards, 155; practitioner listings, 76–77; view of medical community, 143

Christian Science Monitor, 11, 108, 191, 194–95

Christian Science Practitioners: advertisements for, 49–51, 65; court trials of, 150–59, 178; definition of, 45, 66; establishing a practice, 48–52, 61; fees ("compensation"), 59, 72–75, 194; female dominance among, 37; income, 72–73; numbers of, 34, 115, 194, 209; patient numbers, 50–51; professionalization and oversight, 158–59; public reception of, 48, 51; relationship to doctors, 176–78; reputation of, 76–77; social and economic advantages for, 41–45, 60; specialization among, 73, 158; titles of, 49, 96, 153, 154, 159; training of, 45–48, 96, 139

Christian Science Publication Committee, 158

Christian Science Publishing Society, 11, 108, 189, 192

Christian Science sanatoriums, 195

Christian Science Sentinel, 108, 189

Christian Science Theological Seminary, 90

Christian Science trials, 146

Christian Scientist Association, 31, 106

Christian Scientists: accommodation to medical legislation, 4, 158–59, 179, 190; and child abuse, 4, 204, 207–8; and contagious disease, 4, 80, 168–69, 174, 182–83; conversion of, 33–34, 35, 38; defining orthodoxy, 92–94; and Eddy's authority, 85–88, 106–9; female connection of, 8, 34–44; "generic," 10, 71, 88, 109; legal strategies of, 155–56; and legislation, 3; lives of trial, 8, 11, 12; "pseudo," 10, 85, 87; socioeconomic class of, 42–43, 122; as

teachers, 52–53; "true," 10, 85, 87; and vaccination, 80, 179; view of medical theory, 172–73

Church Manual, 11, 108, 158, 159, 209

Church of Christ, Scientist, 7, 31, 87, 106

Church of the Divine Unity, 89

Cincinnati Enquirer, 138, 139, 140

Clark, C. A., 78

Coate, Lloyd B., 61

Coffin, J. L., 83

Cole, Willis Vernon, 61, 65, 68, 69, 165

Collson, Mary, 38, 40, 42, 43

Committee(s) on Publication (COPs), 4, 68, 99, 101, 108, 123, 177, 201; effectiveness of, 161–62; purpose for, 161; strategies of legislative "exemption" and "exclusion," 198–202

Contagious disease: catch-22 for Christian Scientists, 174, 182–83; Christian Scientists barred from treating, 182; legal requirement to report, 80; and public health, 168–69

Cooper, Frank, 163

Corey, Arthur, 65

Coriat, Isador H., 135

Cornelius, Mr. and Mrs. Edward, 201

Corner, Abby H., 10, 82–85, 91, 99

Cotton, J. M., 180

Crafts, Hiram, 24

Craig, Henry K., 117

Cramer, Malinda E., 91

Cross, J. C., 82

Cross, Sarah, 101

Crummer, B. F., 154

Cullis, Charles, 84

Davenport Sunday Democrat, 170

Davis, Emma S., 50, 61–62, 64, 78

Defence of Christian Science, 93

Dent, Frank, 147

Dent v. West Virginia (1889), 147

Dickey, A. H., 163

Dickson, William B., 176

Diehl, Mrs. Andrew, 96, 97